Praise for *Seeds of Deception*

"*Seeds of Deception* is a major event in informing the public about the safety or (more precisely the lack of it) of genetically modified foods, which are hailed to be one of the most important scientific developments of our age. In contrast to the bland assurances from official propaganda, the book lays bare the concerted machinations of the biotechnology industry, the media, politicians and the regulatory authorities, all united in their effort and using any means to allay the rightful concerns and fears of the public about this unpredictable and unsafe technology. A particular strength of the book—and this will be hated by the pro-GM lobby—is that it uses a very colourful but easily understandable language to describe what is usually regarded as 'high' science. My greatest compliment is that even though I am a scientist I got some special insights into the workings of the recombinant DNA technology from Jeffrey Smith's enjoyable presentation."

—Arpad Pusztai, Ph.D.
Leading Expert on Safety Research Conducted on GM Foods

"Jeffrey Smith's lucid, informative and tightly argued exposé of genetically modified foods lays bare the blockbuster food safety issue of the 21st century. Although Americans slept as the biotech industry quietly kidnapped our food supply, Europe sent the miscreants—typified by Monsanto—packing. Mr. Smith presents a roadmap of insensitivity by America's food giants and a blueprint of action to stop them from wreaking further havoc."

—James S. Turner
Author, *The Chemical Feast:*
The Nader Report on the Food and Drug Administration

"To get 'up to speed' on the subject of genetically modified foods, there's no better book than *Seeds of Deception*. Jeffrey Smith deals with the science as well as the deception in a pleasing, sure-footed way."

—Jim Diamond, M.D.
Chair, Sierra Club Genetic Engineering Committee

"In a highly accessible story format, *Seeds of Deception* provides a compelling and powerful indictment of industry and government agencies that have conspired to hide the dangers of genetically engineered foods. After reading the truth about these foods, consumers will avoid eating them; food company executives will want them out of their products; and the government, hopefully, will put an end to this outrageous experiment."

—**Larry Bohlen**
Director of Health and Environment Programs,
Friends of the Earth USA

"*Seeds of Deception* is the first book to make a convincing case for the existence of a genuine conspiracy on the part of the biotechnology industry to suppress free speech, debate and even scientific dialogue about the safety and value of GMOs. In doing so, Jeffrey Smith paints a vivid and disturbing picture of governmental passivity and scientific neglect of urgent problems associated with genetically engineered agriculture. By putting together over a dozen episodes of interference and collusion against activists who have questioned the wisdom of proceeding unabated with this collective, non-consensual experiment with our food, Smith shows how industry proponents have done themselves and a whole generation of consumers a massive disservice in the name of corporate profits and short-term private gain."

—**Marc Lappé, Ph.D.**
Co-Director, The Center for Ethics and Toxics (CETOS)

"There is mounting evidence that genetically modified foods are unsafe. This book, which is the best written on the subject, is essential reading for food activists and concerned consumers."

—**Ronnie Cummins**
National Director, Organic Consumer's Association, USA
Co-author, *Genetically Engineered Food:
A Self-Defense Guide for Consumers*

"This pivotal exposé leaves no doubt that politics and corporate influence, not sound science, allowed these potentially dangerous GM foods onto supermarket shelves."

—Joe Mendelson
Legal Director, Center for Food Safety, USA

"I have seen first-hand how Monsanto and the FDA have resorted to scientific deceit of the highest order to market genetically engineered (rbGH) milk. With captivating style and a flair for describing science in clear, accurate language, *Seeds of Deception* unveils the distortions, omissions and lies for all to see. It is a powerful antidote to this global charade."

—Samuel S. Epstein, M.D.
Professor Emeritus Environmental and Occupational Medicine,
University of Illinois at Chicago School of Public Health;
Chairman, Cancer Prevention Coalition

"The revelations in this book are being made public at a pivotal time in the global GMO debate, and could tip the scales against the biotech industry. The evidence refutes US science and safety claims, and undermines the basis of their WTO challenge. It also presents a compelling argument that nations may use to ban GM foods altogether."

—Andrew Kimbrell
Director, Center for Food Safety, USA

IS THERE A DIFFERENCE BETWEEN O AND NON O FOOD.
If you take a bucket of herbicide, insecticide and
fungicide is it safe? No. How much vegetable do
have to add to make it safe?

Seeds
of
Deception

**EXPOSING CORPORATE AND
GOVERNMENT LIES ABOUT THE
SAFETY OF GENETICALLY
ENGINEERED FOOD**

Jeffrey M. Smith

Foreword by Michael Meacher

Peter,

Safe eating

Jeffrey Smith

Green Books

First published in the UK in 2004
by Green Books Ltd,
Foxhole, Dartington,
Totnes, Devon TQ9 6EB

Originally published in the USA in 2003 by
Yes! Books, P. O. Box 469, Fairfield, IA 52556
(888) 717 7000
www.seedsofdeception.com

Cover design by Rick Lawrence
samskara@onetel.net.uk
adapted from the design for the US edition by George Foster

ISBN 1 903998 41 7

Printed on Five Seasons Stone-White 100% recycled paper
by MPG Books, Bodmin, Cornwall, UK

Trademark Acknowledgements

StarLink® is a registered trademark of Aventis Crop Science.
Roundup® and Roundup Ready® are
registered trademarks of Monsanto Company.
NutraSweet® is a registered trademark of NutraSweet Property Holding, Inc.
Quest® is a registered trademark of Vector Tobacco Ltd. Company.

Contents

11/07/04 Swedish Kenth to brew light
lager containing Bt corn with funding from
Monsanto, DuPont, Bayer CropScience, Syngenta.

Acknowledgements

I deeply appreciate the contributions that so many
people made to this book—as reviewers, as sources,
as moral support. I name only a few here.
Andrea Smith, Rick Smith, Morton Smith,
Nancy Tarascio, Robynn Smith, Arpad Pusztai, Ph.D.,
Michael Hansen, Ph.D., Brian Stains, Bill Crist,
Steve Druker, Barbara Keeler, Helen Whybrow,
Margo Baldwin, Pete Hardin, Joe Cummins, Ph.D.,
Robert Roth, Jane Akre, Steve Wilson,
Ignacio Chapela, Ph.D., James Turner,
Bill Freese, Betty Hoffing, Barbara Reed Stitt,
Gerald Gleich, M.D., Phillip Hertzman, M.D., Rick North,
Ronnie Cummins, Jeff Peckman, Joe Mendelson,
Craig Winter, Bill Lashmett, Howard Vlieger,
Samuel S. Epstein, M.D., David Schubert, Ph.D.,
Robert Cohen, Larry Bohlen, Dick Kaynor,
Britt Bailey, John Kremer and Carol Kline.

Foreword
by Michael Meacher

This is a brilliant book which combines shrewd dissection of the true nature of GM technology, a devastating critique of the health and environmental hazards of GM crops, and scarifying examples of the manipulation of both science and the media by the biotech industry.

Despite the British Government's GM Nation Debate in mid-2003, the level of understanding of GM remains alarmingly low in the UK. This book should be compulsory reading, not only for the general public, but even more so for the decision-makers who have never been exposed to systematic analysis of the problems created by GM.

What is so exciting about this book is that it is no dry text of scientific exegesis—it positively fizzes with the human drama of the cabals and conspiracies behind the scenes which have littered the history of Big Biotech in its frantic efforts to get itself accepted. It is meticulously documented and powerfully written, somewhere between a documentary and a thriller.

It reveals above all that GM is not some arcane issue about science or technology—it is ultimately about power. There are no consumer benefits from GM crops, the alleged benefits to farmers are deeply disputed, environmental and health testing has never been carried out, non-GM farmers are being put seriously at risk. So why is GM being pressed at all? The answer, set out painstakingly and frighteningly in this book, tells us a great deal about how power is exercised today—funding political parties and key individuals, networking around opinion-formers and decision-makers, and fixing strategic job swaps between the biotech industry and Government. And this is not just conjecture; plenty of examples are given which illustrate how secretive and malign these influences are.

The main area of cover-up is undoubtedly the GM effects on health. It is a staggering fact that there have been virtually no clinical or biochemical tests of the impacts of eating GM foods on human health. Jeffrey Smith sets out, like a detective story, the unravelling of the L-tryptophan fiasco, the StarLink maize allergy mishap, and the cauliflower mosaic virus promoter hazard, as well as a host of other health risks, both predicted and unpredictable.

But the kernel of the book is the commercialization of politics and the politicization of science. For those who still believe the constitutional fantasy that governments act in accordance with their manifesto in the general interests of society, this book will come as a shocker. The exercise of power today is much more hard-nosed and ruthless, and the power-brokers are not the electorate, but Big Business. As a case study of this suborning of democratic accountability, Jeffrey Smith's account is an eye-opener. But most of all it is a call to arms, not only to prevent the contamination of the nation's food supply, but even more to tackle the poisoning of the nation's decision-making system by the undercover wielding of economic and financial muscle and PR manipulativeness of Big Biotech.

March 2004

Rt. Hon. Michael Meacher MP was Minister of State for the Environment from 1997 to June 2003. He has served longer as a minister and opposition spokesman than anyone in the current government, having been on the front bench for 27 of the past 30 years since Harold Wilson appointed him a junior industry minister in 1974.

Introduction

On May 23, 2003, President Bush proposed an Initiative to End Hunger in Africa using genetically modified (GM) foods. He also blamed Europe's "unfounded, unscientific fears" of these foods for hindering efforts to end hunger. Bush was convinced that GM foods held the key to greater yields, to expanded US exports, and to a better world. His rhetoric was not new. It had been passed down in the US from president to president, and delivered to the American people through regular news reports and industry advertisements.

The message was part of a master plan that had been crafted by corporations determined to control the world's food supply. This was made clear at a biotech industry conference in January 1999, where a representative from Arthur Andersen Consulting Group explained how his company had helped Monsanto create that plan. First, they asked Monsanto what their ideal future looked like in fifteen to twenty years. Monsanto executives described a world with 100 percent of all commercial seeds genetically modified and patented. Andersen Consulting then worked backward from that goal, and developed the strategy and tactics to achieve it. They presented Monsanto with the steps and procedures needed to obtain a place of industry dominance in a world in which natural seeds were virtually extinct.

Integral to the plan was Monsanto's influence in governments, whose role was to promote the technology worldwide and to help get the foods into the marketplace quickly, before resistance could get in the way. A biotech consultant later said, "The hope of the industry is that over time, the market is so flooded that there's nothing you can do about it. You just sort of surrender."[1]

The anticipated pace of conquest was revealed by a conference speaker from another biotech company. He showed graphs projecting the year-by-year decrease of natural seeds, estimating that in five years, about 95 percent of all seeds would be genetically modified.

While some audience members were appalled at what they judged to be an arrogant and dangerous disrespect for nature, to the industry this was good business. Their attitude was illustrated in an excerpt from one of Monsanto's advertisements: "So you see, there really isn't much difference between foods made by Mother Nature and those made by man. What's artificial is the line drawn between them."[2]

To implement their strategy, the biotech companies needed to control the seeds—so they went on a buying spree, taking possession of about 23 percent of the world's seed companies. Monsanto did achieve the dominant position, capturing 91 percent of the GM food market. But the industry has not met their projections of converting the natural seed supply. Citizens around the world, who do not share the industry's conviction that these foods are safe or better, have not "just sort of surrendered".

Widespread resistance to GM foods has resulted in a global showdown. US exports of genetically modified maize and soya are down, and hungry African nations won't even accept the crops as food aid. The EU is implementing a more stringent labelling and traceability programme. Monsanto is faltering financially and is desperate to open new markets. The US government is convinced that the European Union's (EU) resistance is the primary obstacle and is determined to change that. On May 13, 2003, the US filed a challenge with the World Trade Organization (WTO), charging that the EU's restrictive policy on GM food violates international agreements.

On the day the challenge was filed, US Trade Representative Robert Zoellick declared, "Overwhelming scientific research shows that biotech foods are safe and healthy." This has been industry's chant from the start, and is the key assumption at the basis of their master plan, the WTO challenge, and the president's campaign to end hunger. It is also, however, untrue.

The following chapters reveal that it is industry influence, not sound science, which has allowed these foods onto the market. Moreover, if overwhelming scientific research suggests anything, it is that the foods should never have been approved.

Just as the magnitude of the industry's plan was breathtaking, so too are the distortions and cover-ups. While many of the stories in this book reveal government and corporate manoeuvring worthy of an adventure novel, the impact of GM foods is personal. Many people in the world eat them at every meal. These chapters not only dismantle the position that the foods are safe, they inform you of the steps you can take to protect yourself and your family.

February 2004

Chapter 1

Suppressing the Evidence

When Susan answered the door, she was startled to see several reporters standing in front of her. More were running from their cars in her direction and she could see other cars and TV news vans parking along her street.

"But you all know that we can't speak about what happened. We would be sued and—"[1]

"It's OK now," the reporter from Channel Four Television interrupted, waving a paper in front of her. "They've released your husband. He can talk to us."

Susan took the paper.

"Arpad, come here," she called to her husband.

Arpad Pusztai, a distinguished looking man in his late sixties, was already on his way. As his wife showed him the document, the reporters slipped past them into the house. But Arpad didn't notice; he was staring at the paper his wife had just handed him.

He recognized the letterhead at once—The Rowett Institute, Aberdeen, Scotland. It was one of the world's leading nutritional institutes and his employer for the previous thirty-five years—until his sudden suspension seven months ago. And there it was, clearly spelled out. They had released their gag order. He *could* speak.

The document was dated that same day, February 16, 1999. In fact, less than twenty minutes before, thirty reporters had sat in the Rowett Institute press conference listening to its director, Professor Philip James, casually mention that the restrictions on Dr. Pusztai's speaking to the press had been lifted. Before James had finished his sentence, the reporters leaped for the door. They jumped into their cars and headed straight to the Pusztai's house on Ashley Park North, an address most were familiar with, having virtually camped out there seven months earlier. Now those thirty reporters, with TV cameras and tape recorders, were piled into the Pusztai's living room.

Arpad Pusztai read the document—twice. As he looked up, the reporters started asking him questions all at once. He smiled, and breathed more easily than he had in a long time. He had all but given up hope. Now he finally had the chance to share what he knew about the dangers of genetically engineered foods.

The story of Arpad Pusztai made headlines throughout Europe for months, alerting readers to some of the serious health risks of genetically modified (GM) foods. It was barely mentioned, however, in the US press; the media watchdog group Project Censored described it as one of the ten most underreported events of the year.[2] In fact, major US media avoided almost any discussion of the controversy over genetically modified organisms (GMOs) until May 1999. But that was all about saving the monarch butterfly from GM maize pollen, not about human food safety.

It wasn't until the massive food recall prompted by StarLink®* maize that Americans were even alerted to the fact that they were eating GM foods everyday. Moreover, the American press was forced to question whether GM foods were safe. Up until then, the media had portrayed European resistance to America's GM crops as unscientific anti-Americanism. But as the story of Arpad Pusztai reveals, the European anti-GMO sentiment had been fuelled, in part, by far greater health risks than the scattered allergic reactions attributed to StarLink.

First Shock

Arpad Pusztai was more than good at his work. In other professions, they would call him great. But in the conservative and exacting world of experimental biology, the accolade given was "thorough". Pusztai's thoroughness over fifty years had put him at the top of his field. He had published nearly 300 scientific articles, authored or edited twelve books, and regularly collaborated with other leading researchers around the globe.

In 1995, Arpad, his wife Susan—also a distinguished senior scientist—and colleagues at the Rowett Institute, Scottish Crop Research Institute and University of Durham School of Biology were awarded a £1.6 million research grant by the Scottish Agriculture, Environment and Fisheries Department. Selected over twenty-seven other contenders, this consortium of scientists, with Arpad Pusztai as their coordinator, was chosen to create a model for testing genetically modified (GM) foods, verifying that

---

* StarLink is a registered trade mark of Aventis.

they were safe to eat. Their testing methods were to become the standard used in Britain and likely adopted throughout the European Union.

At the time of the grant, no research had yet been published on the safety of GM foods, and the world's scientific community had plenty of questions and concerns. Pusztai and his team, therefore, were charged with designing a testing regimen that would create confidence and, of course, be *thorough*.

The team's research had been underway for about two years when, in April 1998, the Rowett Institute's director, Professor Philip James, walked into Pusztai's office and placed a sizable stack of documents on his desk. He called in Susan from the adjoining office.

He told the Pusztais that ministers from throughout Europe were about to meet in Brussels to cast their votes regarding regulation of genetically engineered foods. The documents were submissions from biotech companies seeking approval of their own varieties of GM soya, maize and tomatoes. The British Ministry of Agriculture, Forestry and Fisheries (MAFF) was attending the conference and needed a scientific basis with which to recommend them.

Professor James was one of twelve scientists who comprised the Advisory Committee on Novel Foods and Processes (ACNFP), which was responsible for evaluating GM foods for sale in Britain. James was in charge of the nutritional analysis.

Pusztai looked at the stack of papers. There were about six or seven folders, each representing a different request for approval—nearly 700 pages in all. Pusztai knew that James and the other eleven ACNFP members would never actually read these documents themselves. They were extremely busy men. Professor James, for example, served on about a dozen such committees and spoke regularly at international conferences. He was away from the institute so often that Pusztai would frequently greet him in the halls with, "Hello stranger." Besides, James and most of the others were not active scientists. They were committee men—involved in raising money, setting policy, and looking after the politics of science. Arpad and Susan, on the other hand, had already been working for more than two years on designing the methods for approving GM foods. And as part of their grant, they were conducting tests on a new variety of genetically engineered potatoes that the Scottish Ministry had hopes of commercializing. They didn't just know the theory; they had practical experience. The Pusztais were therefore among the most qualified scientists in the world to read and evaluate the stack James had just handed to them.

"How soon does the minister need his recommendations?" asked Pusztai.

"Two and a half hours," said James.

Arpad and Susan quickly got to work. They divided the submissions and focused right in on the most substantial evidence in the documents—the research design and the data.

As Arpad Pusztai looked first at one submission and then another, he was flabbergasted.

"As a scientist, I was really shocked," Pusztai said. "This was the first time I realized what flimsy evidence was being presented to the committee. There was missing data, poor research design and very superficial tests indeed. Theirs was a very unconvincing case. And some of the work was really very poorly done. I want to impress on you, it was a real shock."

Whereas Arpad and Susan had originally thought that two and a half hours would be enough only to give the minister preliminary recommendations on the submissions, it turned out to be more than enough time to give him an answer with confidence. The research presented was in no way adequate to demonstrate that the genetically modified foods described were safe for human or animal consumption. All of them failed to produce sufficient evidence. Pusztai made the phone call.

"I told the minister, on the basis of what we had seen so far, even with just two and a half hours of review, I advised him to be extremely cautious and not accept it," said Pusztai. "And then he said something on the phone which I found really amazing: 'I don't know why you are telling me this, Professor James has already accepted it.'"

Pusztai was stunned. It turned out that not only had the committee approved the GM food submissions based on flimsy evidence, the approvals had taken place two years earlier—James had only wanted some scientific assurances for the minister to use. And neither Pusztai, nor other scientists working in the field, nor the more than 58 million people in the UK knew that they were already eating GM tomatoes, soya and maize—and had been for almost two years. The approvals had all been done under the cloak of secrecy.

The incident was a turning point for Pusztai. Up until then, he had been confident that the scientific and regulatory community would carefully and thoroughly scrutinize this new technology. But now he was concerned. Very concerned.

After the call, Pusztai talked to Professor James and told him why he thought the committee's approval of the foods was a mistake. He said that there were critical pieces of evidence missing and described how the model that his team had developed with their own research was many, many times more rigorous and detailed than what was presented by the biotech companies. Already he was seeing some evidence of dangers in the potatoes he was studying that would not have been picked up in the superficial research done on GM tomatoes, maize and soya.

Professor James was not defensive of the committee's decision. In fact, he was supportive of Pusztai's conclusions, even enthusiastic. If scientists at his institute had created a better way to test GM foods, he reasoned, this could result in very lucrative contracts—millions of pounds pouring in.

"He thought it was a good opportunity to get more funds for scientific research," said Pusztai. "You understand, we are all strapped for cash, all academics. He thought that we should carry on with this research and come up with really great things."

Pusztai, on the other hand, was not enthusiastic. He had serious concerns about the untested GM tomatoes, soya and maize being sold in grocery stores. This was compounded by the fact that he knew that soya, maize and their derivatives are found in about 70 percent of all processed foods.

As Pusztai continued his research, his concerns about GM food intensified.

Hot Potatoes

Pusztai's consortium of scientists was altering the DNA of a potato so that it would do something no potato had ever done before. It was to produce its own pesticide, a lectin, normally found in the snowdrop plant that protects it from aphids and other insects. The industry's goal was to mass produce this combination potato/insecticide, relieving farmers of the burden of having to spray the fields themselves. As part of the research, Pusztai and the team at the Rowett were to test the potato's effects on the health of rats.

Genetically modified potatoes were already being sold and consumed in the United States. Their DNA was spliced with a gene from a soil bacterium similar to *Bacillus anthrax*. The added gene caused the potatoes to create their own pesticide called *Bacillus thuringiensis* toxin or Bt. If insects had the misfortune to eat one of these genetically modified wonders, the Bt, which was manufactured by every cell of the plant, quickly killed the insect. The same Bt-creating genes have also been placed into the DNA of maize and cottonseed, also sold and consumed in the United States, and all officially classified as pesticides by the US Environmental Protection Agency. However, the United States Food and Drug Administration (FDA) had made it clear that in their view, genetically modified crops were assumed to be safe and to offer similar nutritional value as their natural counterparts. This assumption is the cornerstone in US policy, allowing millions of acres of GM food to be planted, sold and eaten without prior safety testing.

Pusztai's team engineered a potato plant to create a different pesticide—a lectin, a natural insecticidal poison that some plants produce to ward off insects. Arpad Pusztai had spent nearly seven years researching this lectin's properties. He was the world's expert on lectins and he knew this particular lectin was safe for humans to eat. In fact, in one of his published studies, he fed rats the equivalent of 800 times the amount of lectins that the GM potatoes were engineered to produce, with no apparent damage. So when he fed the rats his lectin-producing potatoes, Pusztai didn't expect any problems.

What Pusztai and his team found was quite a shock. First, the nutritional content of some GM potatoes were considerably different from their non-GM parent lines, even though they were grown in identical conditions. One GM potato line, for example, contained 20 percent less protein than its own parent line. Second, even the nutritional content of sibling GM potatoes, offspring of the same parent grown in identical conditions, was significantly different.

If Pusztai's results were limited to just these facts, they alone might have undermined the entire regulatory process of GM foods. FDA policy was based on the assumption that genetically modified foods were stable. Nutrient levels were not supposed to vary.

But these findings were completely eclipsed by Pusztai's other, more disturbing discoveries. He found that rats which were fed GM potatoes suffered damaged immune systems. Their white blood cells responded much more sluggishly than those fed a non-GM diet, leaving them more vulnerable to infection and disease. Organs related to the immune system, the thymus and spleen, showed some damage as well. Compared to rats fed a non-GM control diet, some of the GM-fed rats had smaller, less developed brains, livers and testicles. Other rats had enlarged tissues, including the pancreas and intestines. Some showed partial atrophy of the liver. What's more, significant structural changes and a proliferation of cells in the stomach and intestines of GM-fed rats may have signaled an increased potential for cancer.

The rats developed these serious health effects after only ten days. Some of these changes persisted after 110 days, a time period corresponding to about 10 years of human life.

In preparing the diet, Pusztai had been characteristically thorough. Comparisons had been made between rats fed GM potatoes, natural potatoes, and natural potatoes spiked with the same amount of pure lectin as found in the GM potato. The researchers varied the potato preparation, using raw, boiled and baked potatoes, and varied their amounts in the diet. They also varied the total protein content of the diets and tested all these variations over both 10-day and 110-day periods. These testing protocols had all been thoroughly scrutinized and approved in advance by the government's funding office and were consistent with several published studies.

In the end only the rats that ate the GM potatoes suffered the serious negative effects. From the evidence, it was clear that the lectins were not the major cause of the health damage. Rather, there was some effect from the process of genetic engineering itself that caused the damaged organs and immune dysfunction of the adolescent rats. "We used exactly the same methods of genetic engineering as used by the food companies," says Pusztai.

Pusztai knew that his results strongly suggested that the GM foods already approved and being eaten by hundreds of millions of people every day might be creating similar health problems in people, especially in children.

Pusztai was in a terrible bind. He knew that if his potatoes had been subjected to the same superficial studies and approval process that the GM tomatoes, soya and maize had, they too would have flown through the ACNFP approval process without a hitch. They would have ended up on supermarket shelves and in frying pans worldwide.

And Pusztai knew that the superficial research that had been done on the GM tomatoes, soya and maize would not have picked up the types of serious problems he encountered. Furthermore, if human beings developed problems similar to his rats, it could take years to appear and it would be highly unlikely for anyone to suspect GM foods as the cause.

"I had facts that indicated to me there were serious problems with transgenic food," said Pusztai. "It can take two to three years to get sci-

ence papers published and these foods were already on the shelves without rigorous biological testing similar to that of our GM potato work."[3]
If he waited that long, he thought, who knows what kind of serious damage might be inflicted on unsuspecting consumers.

As Arpad Pusztai contemplated these ramifications and compiled his
findings for publication, he was approached by the TV programme
"World in Action". They were anxious to air a scientist's opinion on the
safety of genetically modified foods and were particularly keen to hear
from Pusztai. They knew that his team was the only one in the world conducting thorough feeding trials on GM foods.

Their request brought Pusztai's conflict to a head. The traditional code
of practice of a scientist dictates that he remain silent about his findings
until he can present them at a conference or via publication. But his codes
of ethics dictated that he warn the public immediately about his findings.

Pusztai was also encouraged to speak out by the fact that the research
was publicly funded. "The British taxpayer has spent £1.6 million for this
Rowett-based research. [They] have paid for it," he said. He also knew
that the interview would only provide time for a two- to three-minute
summary. It would therefore not pre-empt the more detailed disclosure
that would come with publication.[3]

He sought the permission of James, who was encouraging. They both
agreed, however, that Pusztai should not be forthcoming with the details
of the data, as that would be more appropriately first published in his
research paper. James had the Rowett Institute's public relations officer
join Pusztai at the studio for the taping.

Pusztai's interview lasted about two hours and was eventually edited
for a 150-second broadcast. The final cut included Pusztai saying that the
effect of the experimental GM potatoes on rats "was slight growth retardation and an effect on the immune system. One of the genetically modified potatoes, after 110 days, made rats less responsive to immune effects."

Asked if he would eat GM foods himself, he said, "If I had the choice I
would certainly not eat it till I see at least comparable experimental evidence
which we are producing for our genetically modified potatoes. I actually
believe that this technology can be made to work for us. And if genetically
modified food will be shown to be safe then we have really done a great
service to all our fellow citizens. And I very strongly believe in this, and
that's one of the main reasons why I demand to tighten up the rules, tighten
up the standards."

He added, "We are assured that: 'This is absolutely safe. We can eat it all the time. We must eat it all the time. There is no conceivable harm which can come to us.' But, as a scientist looking at it, actively working on the field, I find that it is very, very unfair to use our fellow citizens as guinea pigs. We have to find the guinea pigs in the laboratory."[4]

Pusztai was aware that his comments would cause a stir, but he never imagined the magnitude of the controversy that it created.

Eruption in the Media

In contrast to the nearly complete lack of information on GMOs in the United States, the controversy was already running at a low boil in the UK. Monsanto Corporation, the biotech giant, was running full-page advertisements in newspapers touting the benefits of GM foods and attempting to enlist a skeptical public. Major newspapers, on the other hand, were running articles and editorials accusing these ads of misleading the public with false statements. Scientists were quoted in the papers expressing doubts about the foods' safety. And the public was already reeling from the impact of mad cow disease, which had killed several people in spite of the government's earlier assurances of safety. Onto this fertile field dropped Pusztai's bombshell.

On Sunday, August 9, 1998, the day before the airing of "World in Action", the station broadcast advertisements of the interview, highlighting some of Pusztai's points and urging listeners to tune in the next day. At midnight, the station sent a news release throughout the British press. Some reporters started calling immediately, keeping Pusztai up until the early hours.

When Pusztai arrived at work, "the Institute was already bombarded with all sorts of questions from the press and from the Ministry of Agriculture and Fisheries in London, who hadn't been told about the interview," says Pusztai.

But by late morning, the phones fell silent. Pusztai initially figured that the flurry was over, and he could go back to his work. "I found out later that I was sort of made to shut up by eleven in the morning," recalls Pusztai. "The director took over everything, all the PR work, by switching my phone in my office to his office and intercepting faxes and emails; so much so that even our son couldn't get in touch with us."

Professor James, in the meantime, was enjoying unprecedented popularity. "He was on TV every ten minutes or so. He gave his interpretation of how great this work was—a huge advance in science." Pusztai recalls, "He tried to milk every drop of that publicity." Professor James even issued a press release that morning about the team's research, without discussing it or checking it with Pusztai. *For further information*, it said, *contact Dr. Philip James.*

"He thought this was a great thing and he's going to be world famous for it," said Pusztai.

James had reason to have big aspirations. Tony Blair had asked James to draw up the blueprint for a new Food Standards Agency—a kind of British version of the US FDA, only dealing exclusively with food. This was to be a prestigious agency, staffed by 3,000 civil servants. And everyone assumed that Professor Philip James Ph.D. was to be its first director—a significant political appointment.

Now it appeared that James was intent on adding another feather to his cap and perhaps impressing his future boss Tony Blair. So James commandeered the publicity and started giving out the information about the potato research himself.

The problem was—he was wrong. The information he gave to the press, wrote in the release and spoke about on TV was incorrect. He hadn't bothered to check his facts with Pusztai or any member of his team.

Most critical among his mistakes was the type of lectin the research team had used. They had engineered a potato to produce a lectin from the snowdrop plant, called GNA, known to be completely harmless to rats and humans. The lectin James described, however, was "concanavalin A"—a well-known toxic immune suppressant.

His mistake completely misled the public. If the rats were damaged by an experimental potato that was genetically engineered to produce a known toxin, so what? Press reports acknowledged that's what toxins do—so what's the problem? The potato was not on the grocery shelves and never would be.

But Pusztai's lectin was harmless. James' mistake, therefore, sidestepped the bigger issue—the damage to the rats did not come from the lectin, but apparently from the same process of genetic engineering that is used to create the GM food everyone was already eating.

By Monday afternoon, Arpad and Susan figured out that their phones had been re-routed so James could handle the press himself. And by that evening's broadcast, they realized that James had been feeding the press the wrong story. On Tuesday the scientists made several attempts to get to

James, to tell him he was giving out the wrong information to the press. James blocked each attempt. The Pusztais couldn't get to James and no one could get to the Pusztais.

"Our frustration grew with every hour," recounts Pusztai. "All the time he was giving out these press releases and appearing on TV and we could see that he was talking a lot of rubbish. So my wife decided to stop this stupid false misinformation." Susan, with help from the team, wrote down a summary of the actual facts about the research. She limited it to two pages. "We knew if it was a lengthy document he would never read it," says Pusztai.

They were finally able to get a meeting with James at 3 pm on Tuesday. Even though James did not invite Arpad to the meeting, Arpad went along with Susan, their research immunologist, the division head and the deputy director to James' office. Susan handed James the summary.

Everyone became silent as James read the two pages. Pusztai watched the despair surface in James' face when he realized that he had been giving out the wrong information. As James finished the summary, he said softly, "This is the worst day of my life."

"At that point we all agreed that our deputy director, who was very good with words, would make up a much shorter version to be press released on the next morning, so that the controversy would be on a strong scientific foundation. This is how we parted company with Professor James. We were to reconvene the next morning on the twelfth," Pusztai reported.

The next day the Pusztais came to work encouraged that the truth would finally get out. When they were called to a meeting, Arpad Pusztai expected to be handed the corrected release for review. But when he entered the room, the whole top management was assembled. Professor James spoke in a manner that was quite different from that of the previous day. In fact, the Pusztais had never heard him speak that way before.

"He said I was to be suspended, and they will have an audit of the whole business, and then I shall be made to retire," recounts Pusztai. "And my retiring wasn't dependent on what the audit found."

The Institute blocked the team's computers and confiscated all research notes, data, and everything related to the GMO experiments. The research was immediately stopped and the team dismantled.

"This was such an abrupt change in his attitude," says Pusztai. "We parted company before five o'clock in the afternoon on Tuesday and on

Wednesday morning out of the blue I was suspended. This was coming from someone who for two days was milking every ounce of the PR effort, which appeared at the time to be beneficial for him and for the Institute. Something extremely serious must have happened to explain his very sudden and almost 180 degree turn in his opinion and pronouncements."

Pusztai is not sure what prompted this change in Professor James, but he has some ideas. "It was most likely that he had some political interference." In his interviews and releases during the previous two days, James was applauding research that was ultimately critical of the way GM foods on the shelves had been tested. He was also suggesting that more research needed to be conducted (presumably at his Rowett Institute). But, Pusztai points out, "It's no secret that the British Government, particularly Tony Blair, is a supporter of the biotech industry." Pusztai's theory was that James—Blair's primary candidate to head up a major government office— "suddenly blew it. Because for two days he was advocating something which was not the government's policy."

"There are some reports, which are not verified," says Pusztai, "that there were two telephone calls late in the afternoon on the eleventh from Downing Street, from the prime minister's office," forwarded through the Institute's receptionist. (According to the British press, Tony Blair himself had been the recipient of telephone calls from Bill Clinton, who was leaning on Blair to increase support for GM foods.)[5]

Whether it was a directive from the prime minister or some other jolt that prompted James' about-face, suspending Pusztai was clearly an opportunity for James to protect his credibility. If he had released the corrected report and admitted he was giving out false information, his reputation would have been seriously damaged.

Arrows Fly, No Defence

The press was ravenous. "The newspaper men and reporters were almost bedding down on the drive at home," says Pusztai. "I couldn't move out of the house because we were besieged by reporters. The German TV gave hourly updates on the events. I was absolutely blown over by the whole business. I knew that what I did say was not easily accepted. But the reaction to it was absolutely overwhelming."

But Pusztai soon received two threatening letters from Professor James, dated August 18 and 20, which ultimately stopped the press from appearing at the Pusztais' door. "The director invoked my contract which had a prohibition put on me." Pusztai explained that he "could not say anything to anyone without the written permission of the Director."

Pusztai was well aware of the large sums of money that came to the Institute in the form of grants and research contracts. If James claimed that the Rowett Institute lost a project due to Pusztai's statements, he could sue Pusztai for a substantial amount.

"If I say anything to any media person or in fact anybody, I would be taken to court and the Institute would ask for substantial damages from me because I acted against their interest."

"Now I wasn't a very young man," says Pusztai. "I was at the end of my career. I was suspended. I had some savings and my house, which I had worked for all my life. So I wasn't a rich man, and you know how expensive it is to get into litigation. I decided to shut up." His wife, also under contract with the Rowett Institute, was likewise silenced.

With both Pusztais under its gag order, the Institute's PR machine really got rolling. They put out a series of press statements, sometimes contradicting each other, but all designed to discredit Pusztai and his results.

The formal reason given for Dr. Pusztai's suspension was that he had publicly announced the results of his research before they had been reviewed by other scientists, as required by the Rowett Institute. The press was not informed, however, that Director James had enthusiastically given his permission for Pusztai to speak with the press and even called his home after the show to express congratulations. Furthermore, the taping of the show had taken place seven weeks before it aired. If the director had had any second thoughts about airing the show, he had seven weeks to cancel it.

Press statements issued by the Rowett Institute said that results reported by Dr. Pusztai were misleading because he had mixed up the results of different studies. Other statements tried to paint a picture of him as a senile and confused old man or as "muddled" and "on the verge of collapse". James described Pusztai as "absolutely mortified. He is holding his hands up and is apologizing."[6]

Still other statements asserted that the research had not been done on GM potatoes at all, but on a mixture of natural potatoes and lectin. They also indicated that the quality of Dr. Pusztai's research was deficient and claimed that the GM potatoes were not intended to be used as food. A November article in the Institute's publication by the chief executive of the Institute of Biology went even further. He alleged that Pusztai had fabricated findings, "a view he appears to have come to," exclaims Pusztai, "wholly in the absence of seeing any of my working data."[7]

"The Institute thought they could get away with blue murder because they knew I could not reply," says Pusztai. The unchallenged lies about his "mistakes" were sent all over the world and people were led to believe that there was no scientific basis for his warning about GM foods. The *Times* wrote an article: "Scientist's Potato Alert Was False, Laboratory Admits".[8] Another headline from the *Scottish Daily Record and Sunday Mail* read "Doctor's Monster Mistake".[9] Pusztai's credibility and reputation were ruined.

James did not act alone. He handpicked a panel of scientists to conduct an audit of Pusztai's work. It was quite telling that the scientists he selected were not working nutritionists. "That a nutritional institute should select non-nutritionists to do this audit is quite unbelievable," says Pusztai. Moreover, the panel was not given the complete data, did their entire review in less than a day, and didn't consult with Pusztai at all.

A summary of their audit report was released on October 28. It claimed that there were important deficiencies in Pusztai's study. The full audit report, however, was never publicly released. To prevent leaks, only ten copies were printed. Even the chairman of the panel that produced the report was not given a copy.

Loophole

Throughout this period, Pusztai received inquiries from senior scientists around Europe. They had collaborated with him for years and were not fooled by reports of malpractice and senility. They wanted to know the truth. With the threat of a lawsuit by the Rowett over his head, however, Pusztai couldn't tell them what he knew.

But then Pusztai discovered a legal loophole. The contract with the Rowett Institute did not bar him from sharing unpublished research with other scientists. Exchange of information is a long established tradition in scientific circles. Pusztai could, in theory, share his data with these top scientists, provided that it wasn't published.

But a major hurdle remained. "They had confiscated our data," says Pusztai. "I could not use my recollection because science is very precise. If I say something based on my recollection which later turns out not to be absolutely correct, I would truly be destroyed."

In late November he got a break. In response to what had become an enormous media controversy, Parliament asked James to send his evidence against Pusztai for evaluation and to testify before a House of Lords committee. James realized that Pusztai would likely be asked to defend himself and would need his data. James also remembered that the Institute's contract with employees stipulated that when an audit takes place, the "accused" has a legal right to reply to the findings of the audit—again with data in hand. The Institute begrudgingly sent Pusztai some, but not all, of his confiscated data.

Pusztai could now respond to his fellow scientists' requests. He sent them the research design and findings, a copy of the Rowett's audit report and his response to it. The data was compelling. So much so that twenty-three of these scientists from thirteen countries chose to form their own independent panel to conduct a formal peer review and send their report to Parliament.

The panel analyzed Pusztai's data and the Rowett's report. The twenty-three scientists released a memorandum on February 12, 1999, charging that the Rowett's report seemed to select and interpret only those results that would disprove Pusztai's conclusions, while selectively ignoring more relevant data. In spite of this bias, the independent panel said the data analyzed in the audit report nevertheless "showed very clearly that the transgenic GNA-potato had significant effects on immune function and this alone is sufficient to vindicate entirely Dr. Pusztai's statements." They further stated that the data from the audit report combined with Dr. Pusztai's material would in fact be suitable for publication in a peer-reviewed journal. The report stated that "although some of the results are preliminary, they are sufficient to exonerate Dr. Pusztai by showing that the consumption of [GM]-potatoes by rats led to significant differences in organ weight and depression of lymphocyte [immune] responsiveness compared to controls."[10]

The panel of scientists also called for a moratorium on the sale of genetically modified crops.

The controversy was re-ignited in full force. A report published two days later exposed the fact that Monsanto had given the Rowett Institute £140,000 before the blow-up, adding even more fuel to the media's fire.

Under the intense pressure of a highly publicized scandal, Parliament invited Pusztai to present evidence before the Science and Technology Committee of the House of Commons. Parliament's request overrode the Rowett contract—James was forced to release the gag order. That was

February 16; the day the Pusztais unexpectedly hosted thirty members of the press in their living room.

The Battle for Public Opinion

While the European media were hungry for controversial GM food stories, major press in the US had presented them only a couple of times. On October 25, 1998, a cover story in the *New York Times Sunday Magazine* introduced the pesticide-producing Bt potato and how it slipped through the FDA and EPA bureaucracies into the market without thorough safety testing. A major news network ran a GM food story the next week, but then there was nothing for months.

In the UK and parts of Western Europe, however, substantial reporting had led to growing public contempt of GM foods. A leaked October 1998 report prepared by pollster Stan Greenberg for Monsanto said, "The latest survey shows an ongoing collapse of public support for biotechnology and GM foods." Greenberg, who had also conducted opinion polls for President Clinton, Tony Blair and German Chancellor Gerhard Schroeder, wrote, "At each point in this project, we keep thinking that we have had the low point and that public opinion will stabilize, but we apparently have not reached that point. . . . Negative feelings have risen from 38 percent a year ago to 44 percent in May to 51 percent today. A third of the public is now extremely negative, up 20 percent."[11] When the press finally heard the truth about the potato research directly from Pusztai, their reports were destined to push that figure even higher.

The media went wild. That third week in February 1999, more than 1,900 column inches were published about genetic engineering. An editorial declared, "Within a single week the spectre of a food scare has become a full scale war."[12] During the month of February, the British press spewed out more than 700 articles on GMOs.[13] A columnist in *New Statesman* wrote, "The GM controversy has divided society into two warring blocs. All those who see genetically modified food as a scary prospect— 'Frankenstein foods'—are pitted against the defenders."[14]

Among the defenders was the Royal Society, an organization that included many scientists who viewed the attack on GM foods as a threat to their own continued funding and livelihood. On February 23, nineteen fellows of the Society published a letter in *The Daily Telegraph* and the *Guardian* criticizing researchers who "triggered the GM food crisis by publicizing findings that had not been subjected to peer review."

Two weeks later, when Pusztai and James testified before the House of Commons Committee, James also denounced Pusztai for discussing unpublished research. But one Member of Parliament, Dr. Williams, challenged him:

> "There is a real problem for us here, and that is that you say that it is not right to discuss unpublished work; as I understand, all of the evidence taken by the advisory committee [that approves GM foods for human consumption] comes from the commercial companies, all of that is unpublished. This is not democratic, is it? We cannot discuss the evidence because it is not published; there is no published evidence. So we leave it completely to the advisory committee and its good members to take all of these decisions on our behalf, where all of the evidence comes, simply, in good faith, from the commercial companies? There is a hollow democratic deficit here, is there not?" The MP added, "How is the general public out there to decide on the safety of GM foods when nothing is published on the safety of GM foods?"[15]

Although the Labour party in Parliament was quite pro-GMO, members of the Committee also confronted James for issuing public statements about research he knew nothing about and for incorrectly describing the lectin used in the research. Pusztai was further redeemed when James admitted that he had never suspected Pusztai of any wrongdoing or fraud.

In April 1999, the British food industry bowed to consumer pressure. Unilever, Britain's biggest food manufacturer, announced it would remove GM ingredients from its products sold in Europe. "The announcement started a week-long stampede by leading companies, all household names," reported the *Independent*.[16] Nestlé made its announcement the next day, as did the major supermarket chains including Tesco, Sainsbury, Safeway, Asda and Somerfield. McDonalds and Burger King also committed to remove GM soya and maize from their ingredients in European stores. In the end, no major retailer was left standing in the GMO camp. They would eventually spend millions sourcing new supplies of non-GM maize, soya and their derivatives, or re-formulating their recipes, removing maize and soya products altogether. (The European Union passed a law requiring foods that contain ingredients with more than 1 percent GM content to be labelled. Most European producers have eliminated GM ingredients in order to avoid the label. On July 2, 2003, the European Parliament voted to lower the labelling threshold to 0.9 percent.)

Science in the Corporate Interest

With billions at stake, the biotech industry was desperate to contain the anti-GM food rebellion. They needed to do something and fast. But the corporations, particularly Monsanto, couldn't appear to defend themselves directly. "Everybody over here hates us," admitted Dan Verakis, Monsanto's chief European spokesman. He was spotlighted by *The Observer* in a February 21 article entitled: "Food Furore: The Man with the Worst Job in Britain".[17]

Indeed, Norman Baker, a Liberal Democrat Member of Parliament told the House of Commons in March that Monsanto is "public enemy number one". Baker said, "They insist on thwarting consumer choice, bulldozing elected governments and forcing their wretched products on the world's population." He demanded that the corporation's activities be curtailed.[18]

Monsanto and the industry obviously had to work through intermediaries. And according to pollster Greenberg's leaked document, they had friends in high places. The report revealed that Monsanto's strategy was to win over "a socio-economic elite" consisting of Members of Parliament and "upper-level civil servants".[19]

Norman Baker's animosity toward Monsanto, it turns out, was not shared by the leaders of the Labour party. According to a February 1998 report in the *Globe and Mail*, since the party had taken office in the previous year, "government officials and ministers have met companies involved in GM foods eighty-one times (twenty-three with Monsanto alone)." The corporations' efforts paid off handsomely. "More than $22 million has been earmarked in aid for British biotech firms,"[20] and government leaders had been unabashedly pro-biotech.

But now those same leaders were in trouble. Their constituents had become overwhelmingly anti-biotech. According to the minister's own poll, only 35 percent of the British people trusted the government "to make biotechnology decisions on their behalf". The people did not believe that their government would "provide honest and balanced information about biological developments and their regulations". And only 1 percent of the public thought that GM food "was good for society".[21]

The government's credibility on the issue had suffered repeated setbacks. For example, in spite of its claims that GM foods were absolutely safe, a report leaked at the beginning of the year showed that the government wasn't quite sure. The Advisory Committee on Novel Foods and Processes (ACNFP) had been secretly talking with supermarket executives who had access to the food purchasing records of about 30 million customers who used supermarket "loyalty cards". The committee wanted to cross-reference purchasing records with health databases to see if those eating GM foods were more prone to get sick. "The study would specifically look for increases in childhood allergies, cancer, birth defects and hospital admissions."[22] When the report was leaked, the embarrassed government withdrew plans for any monitoring.

Now the government leaders were preparing an initiative to win back public confidence in GM foods. According to a leaked private document obtained by the *Independent on Sunday*, the Health minister, Environment minister and the Food Safety minister met on May 10 and prepared "an astonishingly detailed strategy for spinning and mobilizing support for" GM foods. "One of [the] ministers' main concerns," said the report, "was to rubbish research by Dr. Arpad Pusztai."

The ministers somehow knew in advance that three pro-biotech reports due out in May—by the Royal Society, the House of Commons Committee and the ACNFP—would all attack Pusztai. The ministers therefore planned to have pro-biotech scientists further denounce his work when the reports were released. The scientists, carefully selected by the Office of Science and Technology, would also use the opportunity to "trail the Government's Key Messages", one of which was to convince the public "that industry should be given time to develop and demonstrate possible benefits from GM products." According to the *Independent on Sunday*, many of these so-called "independent" experts, whom the ministers wanted "available for broadcast interviews and to author articles"[23] conveniently "gained their expertise in the pay—direct or indirect—of the [biotech companies]."[24]

In addition, the ministers themselves decided to make numerous media appearances, to "speak with one voice" directly to the people. The Health minister volunteered to write an article that would be published in the prime minister's name. "An instant rebuttal system was to be set up to counter reaction by 'activists and other pressure groups.'" And the ministers were to seek endorsements from the Royal Society and others, which, the document said, "will help us to tell a good story".[23]

It was all to start with the three pro-biotech, anti-Pusztai reports due out in the same week, followed immediately by the ministers' announcement of new programmes related to GM food and a high profile media blitz. It was to be a week to regain consumer confidence.

One of the reports to be made public came from the House of Commons Committee that had heard testimony from Pusztai and James. Pusztai had been confident that they would vindicate him. Although during his testimony, for some reason the MPs did not allow Pusztai to go into the scientific details of his case, he had given them the facts in a document he had carefully prepared over the previous month.

But when the report came out, it selectively omitted or twisted much of his testimony, "flying directly in the face of what actually was said".[7] Indeed, even a cursory comparison between the public transcript of the testimony and the Committee's report shows substantial disparities. It was also clear to Pusztai that they hadn't even read his document. Observers interpreted the Committee's report as the government's attempt to protect the reputation of GM foods, while sacrificing the reputation of Pusztai.

"This was the final straw," confessed Pusztai. When he had escaped communist Hungary as a youth, he chose to relocate in the UK, believing that the people were tolerant and the system was just. But the day he read the report from the highest authority in the land, he says, "My belief in the democratic process was totally shattered."

As the ministers had predicted, Pusztai received a similar rebuff from the Royal Society. The Society had supposedly undertaken a peer review of Pusztai's study. Although the Society does not conduct peer-reviews— not one in their 350-year history—they made an exception. The problem was, they didn't have complete data. They also refused to meet Pusztai, or to reveal the names and qualifications of the scientists who conducted the review. Not surprisingly, their anonymous committee declared Pusztai's work "flawed".[25]

(It was also around this time that Pusztai suffered another indignity— his house was burgled and many of his papers were taken. Soon after, the security camera at the Rowett Institute revealed that a burglar broke into Pusztai's old office, and a burglar was caught breaking into the house of Stanley Ewen, a sympathetic colleague who was following up Pusztai's work.)

After the reports came out, the ministers and their handpicked "independent" scientists did their rounds in the British media. But the week that was designed to regain the public's confidence in GM foods didn't go entirely as planned.

According to the ministers' poll, the public trusted doctors far more than they trusted their government. The ministers were therefore not too pleased when the British Medical Association that same week "called for a moratorium on planting GM crops commercially" and "warned that such food and crops might have a cumulative and irreversible effect on the environment and the food chain."[21] Also that week, it was disclosed that Sir Robert May, the Government's Chief Scientific Adviser said, "the GM crops now being tested should not be approved for commercial use until at least 2003."[24]

Also that week, one of the world's leading medical journals, the *Lancet*, described the Royal Society's unprecedented condemnation of Dr. Pusztai as "a gesture of breathtaking impertinence to the Rowett Institute scientists who should be judged only on the full and final publication of their work." The editorial also said, "It is astounding that the US Food and Drug Administration has not changed their stance on genetically modified food adopted in 1992," which states that they do not believe it is "necessary to conduct comprehensive scientific reviews of foods derived from bioengineered plants". The *Lancet* said, "This stance is taken despite good reasons to believe that specific risks may exist. . . . Governments should never have allowed these products into the food chain without insisting on rigorous testing for effects on health. The companies should have paid greater attention to the possible risks to health." They added, "The population of the USA, where up to 60 percent of processed foods have genetically modified ingredients, seem, as yet, unconcerned."[26]

Researchers at Cornell University announced that same week that monarch butterflies died when they came into contact with pollen from maize engineered to create its own pesticide. This was "the first clear evidence that these crops pose a threat to wildlife."[24] This news shattered the near boycott of coverage on the GMO issue by the US press. Although major media had avoided reporting on GM food safety issues, when the butterfly appeared to be under attack, the press rushed to its aid with months of attention.

And finally that week, the *Independent on Sunday* broke the story about the ministers' secret media plans, which were depicted as the "most damning description yet, of ministers' objectives in the controversy". The paper described the government's actions as a "a cynical public relations exercise", attempting to "save ministers' faces".[23]

The article also said, "The Government has been attacked previously for trying to get sympathetic scientists exposure in the media." In defence, only a week earlier the Agriculture minister Jack Cunningham had assured the paper that "there is no spin-doctoring exercise with scientists" and no attempt to recruit them to "join in some government media campaign". His assurances notwithstanding, the paper now described "secret meetings in which ministers try to spin the issue, even down to trying to fix which 'independent' scientist appeared on the *Today* programme to support the Government line."[24] The paper concluded, "this is the boldest admission so far that [the government] is trying to co-opt [scientists] as part of its PR strategy."

The Royal Society Fights Back

In the coming months, the Royal Society picked up where the ministers left off. According to the *Guardian*, they set up their science policy division in "what appears to be a rebuttal unit". Its purpose "is to mould scientific and public opinion with a pro-biotech line", and to "counter opposing scientists and environmental groups". Among its functions is maintaining "a database of like-minded Royal Society fellows who are updated by email on a daily basis about GM issues".

Rebecca Bowden, who had coordinated the critical peer review of Pusztai, headed the division. Bowden had formerly worked in the government office that regulated GMOs.

Now, in the fall of 1999, her rebuttal unit sprang into action when it learned that the *Lancet* was considering publishing Pusztai's research and had already circulated the paper to six scientists for peer review. Richard Horton, the *Lancet's* editor, told the *Guardian,* "there was intense pressure on the *Lancet* from all quarters, including the Royal Society, to suppress publication."

The paper passed the peer review and was set to appear on October 15, 1999. On October 13, Horton received a call from a senior member of the Royal Society. According to the *Guardian*, Horton, "said the phone call began in a 'very aggressive manner'. He said he was called 'immoral' and accused of publishing Dr. Pusztai's paper which he 'knew to be untrue'. Towards the end of the call Dr. Horton said the caller told him that if he published the Pusztai paper it would 'have implications for his personal position' as editor."

Although Horton declined to name the caller, the *Guardian* "identified him as Peter Lachmann, the former vice-president and biological secretary of the Royal Society and president of the Academy of Medical Sciences."

Lachmann had been one of the nineteen co-signers on the Royal Society's open letter attacking Pusztai. He also had extensive financial ties to the biotech industry: According to the *Guardian*, Lachmann had consulted with the company that markets "the animal cloning technology behind Dolly the sheep", has a directorship on another biotech company, and "is also on the scientific advisory board of the pharmaceutical giant SmithKline Beecham, which invests heavily in biotechnology."[25]

In spite of his threats, the *Lancet* went forward with publication.

Eventually . . . Follow-up Studies Still Not Done

A lot of energy was being spent attacking and defending viewpoints. Very little energy was spent on safety testing.

It would have been fairly straightforward to conduct a follow-up study on Pusztai's research to find out, for example, if any of the GM products we are eating create similar organ or immune system problems. But, having seen what happened to Pusztai, no one was willing to go there.

The British government clearly wasn't. According to one observer from the UK's Natural Law Party, the reason the government had commissioned the research team from the Rowett Institute in the first place was that "it was convinced that it would come up with a favourable result in relation to the safety of the GM potatoes." When Pusztai first discovered the health problems of his rats, even before his TV appearance he requested additional government funding to identify its source. But the government wanted nothing to do with it. In fact, after Pusztai's unexpected discovery, the government ended *all* funding of safety testing.[27]

Pusztai's potato study, plus his earlier paper on experimental GM peas, therefore, remain the only two published independent peer-reviewed feeding studies on the safety of GM foods. As of early 2003, there were only eight other peer-reviewed published feeding studies, all of which were funded directly or indirectly by the biotech companies.

One of these, which has been used by the biotech industry as their primary scientific validation for safety claims, studied the GM soybean called Roundup Ready®*. This soybean is engineered to withstand the normally fatal effects of Monsanto's herbicide called Roundup®*. Using these herbicide-tolerant crops, a farmer can spray his or her field several times during the growing season, making weeding easier. Roundup, which is Monsanto's

---

* Roundup Ready® and Roundup® are registered trade marks of Monsanto Company.

brand name for glyphosate, is the world's best-selling herbicide. Its patent was due to expire in 2000. To prevent a huge loss in market share, Monsanto introduced Roundup Ready crops. Now when farmers buy the GM seeds, they sign a contract requiring them to use only Monsanto's brand, or one of their licensees.

In 1996, Monsanto scientists published a feeding study that purported to test their soybeans' effect on rats, catfish, chicken and cows. But, Pusztai says, "It was obvious that the study had been designed to avoid finding any problems. Everybody in our consortium knew this."[28]

For example, the researchers tested the GM soya on mature animals, not young ones. Young animals use protein to build their muscles, tissues and organs. Problems with GM food could therefore show up in organ and body weight—as it did with Pusztai's young adolescent rats. But adult animals use the protein for tissue renewal and energy. "With a nutritional study on mature animals," says Pusztai, "you would never see any difference in organ weights even if the food turned out to be anti-nutritional. The animals would have to be emaciated or poisoned to show anything."

But even if there were an organ development problem, the study wouldn't have picked it up. That's because the researchers didn't even weigh the organs, "they just looked at them, what they call 'eyeballing'," says Pusztai. "I must have done thousands of post-mortems, so I know that even if there is a difference in organ weights of as much as 25 percent, you wouldn't see it."[3]

Even more troubling was that in a feeding test supposedly designed to detect the effects of GM soya, in one of the trials researchers substituted only one tenth of the natural protein with GM soya protein.[3] In two others, they diluted their GM soya six- and twelve-fold. Scientists Ian Pryme of Norway and Rolf Lembcke of Denmark wrote that the "level of the GM soya was too low and would probably ensure that any possible undesirable GM effects did not occur."

Pryme and Lembcke, who published a paper in *Nutrition and Health* that analyzed all peer-reviewed feeding studies on GM foods, also pointed out that the percentage of protein in the feed used in the Roundup Ready study was "artificially too high". This "would almost certainly mask, or at least effectively reduce, any possible effect of the [GM soya]." They concluded, "It is therefore highly likely that all GM effects would have been diluted out."[29]

The Monsanto paper was "not really up to the normal journal standards," says Pusztai, who had published several studies in that same nutrition journal. In addition to its design flaws, the paper didn't even describe the exact feed composition used in the feeding trials—normally an important journal requirement. "No data were given for most of the parameters," according to Pryme and Lembcke.

The study did, however, reveal several significant differences between Roundup Ready and natural soya in spite of the authors' claims to the contrary. There were significant differences in the ash, fat and carbohydrate content. Roundup Ready soya meal contained "more trypsin inhibitor, a potential allergen".[30] This increase might help explain the sudden jump in soya allergies in the UK beginning right after Roundup Ready soya was introduced. This public health concern is discussed in a later chapter. Also, cows fed GM soya produced milk with a higher fat content, further demonstrating a disparity between the two types of soya.[31]

Researchers measured additional differences between GM and natural soya that, for some reason, were left out of the published paper and were not part of the FDA's review. Years after the study appeared, medical writer Barbara Keeler obtained this missing data from the journal that published the study and broke the story in the *Whole Life Times News*. The omitted information demonstrated that Monsanto's GM soya had significantly lower levels of protein, a fatty acid, and phenylalanine, an essential amino acid. Also, toasted GM soya meal contained nearly twice the amount of a lectin—one that may interfere with the body's ability to assimilate other nutrients. According to Keeler's opinion piece published in the *Los Angeles Times*, the study had several red flags and "should have prompted researchers and the FDA to call for more testing."[30]

Pusztai says that if he had been asked to referee the paper for publication, "it would never have passed."[3] He's confident that even his graduate assistants would have taken the study apart in short order.

According to Michael Hansen of the Consumers Union, the organization that publishes *Consumer Reports* in the US, Pusztai's potato research is "a much better-designed study than the industry-sponsored feeding studies I have seen in peer-reviewed literature that deal with Round-Up Ready soybeans or Bt corn [maize]."[32] A quick review of these is telling.

Two studies looked at GM maize varieties that are currently approved and sold for human and animal consumption. The research was designed for commercial purposes, however, not as safety assessments.

Another maize study employed an experimental variety that was never marketed. It was neither a proper nutritional study nor a safety assessment. Apparently the primary variable used to evaluate the effects of feeding the maize to adult mice, for example, was that the animals did not die.

A Japanese paper attempted to evaluate the effects of Monsanto's Roundup Ready soya on mice and rats, but for some inexplicable reason, researchers used a starvation diet. The young animals gained little or no weight during a very long feeding trial. According to Pusztai, this is equivalent to a child gaining no weight for more than a decade. One possible explanation is that the feed was over-heated and lost its nutritional value. Whatever the reason, no valid conclusions can be drawn from the data.

Besides Pusztai's, there were three additional studies on GM potatoes: One used a potato engineered with a soybean gene. The combination failed to provide the intended increase in protein. The second used a potato engineered with a strong insecticidal toxin. Researchers did not provide a balanced diet to the animals resulting in severe loss of weight and very little usable data. The third looked at potatoes that created their own insecticide using Bt toxin. According to Pusztai, these three were not nutritional studies and the first two were not designed to properly evaluate safety.

The third study, however, did include one important component of a safety assessment—an analysis of tissue samples. Although the authors examined only a small portion of the small intestine, they discovered the same type of unusual increase in cell growth that Pusztai had discovered in the small and large intestine of the rats that ate his GM potato. In fact, that same cell proliferation could explain the increased weight of the cecum and small intestines discovered in Pusztai's earlier study using GM peas. Thus, indications of unusual cell growth in the intestines were found in the only three studies that had the capacity to find it. The implications of this cell growth are unclear, but Pusztai and others say it may be a precursor to cancer.

It is important to note that none of the published studies refuted Pusztai's discovery of damage to organs and the immune system. Similar problems may have afflicted laboratory animals from the other studies, but since the scientists weren't looking for that, their research designs would not have detected them.

One additional unpublished study is worth mentioning. It was conducted on FlavrSavr tomatoes. These tomatoes were genetically engineered to have a prolonged shelf life. As this was the first GM crop to be approved in the US, the manufacturer actually requested the FDA to review their feeding study data—a gesture no subsequent manufacturer has repeated. Documents revealed that many of the rats that ate the GM tomatoes developed lesions in their stomachs. For unknown reasons, researchers did not examine tissue elsewhere in the digestive tract. They also did not provide an explanation as to why seven of the forty rats that were fed the GM tomatoes died unexpectedly within two weeks.

The complete body of research on the safety of GM foods also includes: a study published in a non-peer-reviewed journal, which demonstrated that tissue samples from the digestive tract of both humans and monkeys reacted with GM tomatoes in a test tube[33]; an unpublished feeding study of a GM maize grown in the US, which showed an increased death rate among GM-fed chickens[34]; studies comparing the nutritional content of GM foods with their natural counterparts, demonstrating clear differences between the two types of food; research demonstrating that GM foods can produce new allergens (see Chapter Six); highly controversial studies on the GM bovine growth hormone, which apparently omitted incriminating data (see Chapter Three); and the industry's own studies, such as those submitted to the UK committee that had shocked Pusztai by their inadequacy.

In spite of this small body of research, GM foods are a regular part of the US diet. Approximately 80 percent of the soya and 38 percent of the maize planted in the U.S in 2003 was genetically engineered. Derivatives from these two crops are found in about 70 percent of processed foods. In addition, 70 percent of the cotton crop and more than 60 percent of the oilseed rape (canola) crop, both used for cooking oil, are also genetically modified. About 75 percent of these crops are engineered to withstand otherwise deadly applications of an herbicide, 17 percent produce their own insecticide, and 8 percent are engineered to do both. There are also hundreds of foods produced with genetically engineered cooking agents, food additives and enzymes, as well as varieties of GM squash and papaya. And there are dairy products from cows injected with a GM bovine growth hormone. All these are sold without labels identifying them as GMOs.

The regulations in the US are so lax that there are no required pre-market safety tests. There is no way to determine if these GM foods are creating serious health problems. People get sick all the time without tracking their illness to food, or pesticides, or air or water pollution. The causes remain well hidden.

According to a March 2001 article in the *New York Times*, "The CDC [Center for Disease Control] now says that food is responsible for twice the number of illnesses in the United States as scientists thought just seven years ago. . . . At least 80 percent of food-related illnesses are caused by viruses or other pathogens that scientists cannot even identify."[35] The reported cases include 5,000 deaths, 325,000 hospitalizations, and 76 million illnesses per year. This increase roughly corresponds to the period when Americans have been eating GM food. In addition, obesity has sky-rocketed. In 1990, no state had 15 percent or more of its population in the obese category. By 2001, only one state didn't. Diabetes rose by 33 percent from 1990 to 1998, lymphatic cancers are up, and many other illnesses are on the rise. Is there a connection to GM foods? We have no way of knowing, because no one has looked for one.

Follow the Money

With such slim research on the safety of GM food and such enormous risks, why are respected institutes, scientific panels, research journals, even government officials lining up to defend it as having been proven to be safe? And why are they so quick to condemn evidence that might be used to protect the public? Although subsequent chapters will illustrate how pervasive and dangerous these trends really are, a key to understanding why they happen is to follow the money.

With less research money available from public sources, more and more scientists in the US and Europe are dependent on corporate sponsors, and hence corporate acceptance of their research *and* results. Among Britain's top research universities, for example, dependence on private funds often amounts to 80 to 90 percent of the total research budget.[36] But reliance on corporate sponsorship can carry a hidden price.

A poll of 500 scientists working in either government or recently privatized research institutes in the UK revealed that 30 percent had been asked to change their research conclusions by their sponsoring customer. According to the report, published in the UK's *Times Higher Education Supplement* in September 2000, "The figure included 17 percent who had been asked to change their conclusions to suit the customer's preferred outcome, 10 percent who said they had been asked to do so [in order] to obtain further contracts and three percent who claimed they had been asked to make changes to discourage publication."

If 30 percent admitted to having been asked to change their results, one wonders how many others, having succumbed to their customers' requests, were too embarrassed to answer truthfully.

The article, entitled "Scientists Asked to Fix Results for Backer", said scientists complained that "contracting out and the commercialization of scientific research are threatening standards of impartiality."

Dr. Richard Smith, editor of the *British Medical Journal*, says that the "competing interests" that sponsor research have "quite a profound influence on the conclusions." He warns, "We deceive ourselves if we think science is wholly impartial."[37]

In the US, corporate donations rose from $850 million in 1985 to $4.25 billion in less than ten years. According to the *Atlantic Monthly*, "increasingly the money comes with strings attached. . . . In higher education today corporations not only sponsor a growing amount of research—they frequently dictate the terms under which it is conducted."[38]

Consider the case of the University of California at Berkeley. In November 1998, the biotech company Novartis gave $25 million to the Department of Plant and Microbial Biology for research. In exchange, Novartis gets the first rights to negotiate licences for about one third of the discoveries made by the department. This includes discoveries funded by Novartis as well as those funded by federal and state sources. Novartis can also delay the publication of research by up to four months, providing time for patent applications and for allowing the company to utilize the proprietary information. In addition, Novartis gets representation on two of the five seats of the committee that determines how the department's research money is spent.

When informed of this deal, many in the faculty were outraged. More than half believed it would have a "negative" or "strongly negative" effect on academic freedom, about half thought it would get in the way of "public good research", and 60 percent thought it would inhibit the free exchange of ideas between scientists.

"Worse than the problems of enforced secrecy and delay," says the *Atlantic Monthly* article, "is the possibility that behind closed doors some corporate sponsors are manipulating manuscripts before publication to serve their commercial interests. . . . A study of major research centers in the field of engineering found that 35 percent would allow corporate sponsors to delete information from papers prior to publication."

In addition, many professors own stock in the company that sponsors their research, or sit on their boards, or hold a corporate-endowed position, or simply rely on the corporation for continued research money. Even universities are investing in companies that fund or benefit from university research. "In a study of 800 scientific papers published in a range of academic journals, Sheldon Krimsky, a professor of public policy at Tufts University and a leading authority on conflicts of interest, found that slightly more than a *third* of the authors had a significant financial interest in their reports." None of these papers, however, disclosed the information. Mildred Cho, a senior research scholar at Stanford's Center for Biomedical Ethics, says, "When you have so many scientists on boards of companies or doing sponsored research, you start to wonder, How are these studies being designed? What kinds of research questions are being raised? What kinds aren't being raised?"[38]

Research in the *Journal of the American Medical Association* revealed that studies of cancer drugs funded by non-profit groups were eight times more likely to reach unfavourable conclusions as the studies funded by the pharmaceutical companies. Or consider the case of the genetically modified artificial sweetener aspartame: About 165 peer-reviewed studies were conducted on it by 1995. They were divided almost evenly between those that found no problem and those that raised questions about the sweetener's safety. Of those studies that found no problem, 100 percent were paid for by the manufacturer of the sweetener. All of the studies paid for by non-industry and non-government sources raised questions.[39] The manufacturer of the sweetener, by the way, is GD Searle, which was a wholly owned subsidiary of Monsanto during that period.

Many people agree that the biotech industry has reaped especially great advantage from the academic sector. Sociologist Walter Powell, "believes that the close links between universities and industry are a principal reason why US firms now dominate the biotech market."[38] But, according to University of Minnesota professor Anne Kapuscinski who studies GMOs, that same close link may be making it difficult for scientists to raise questions about GMO safety. This was evidenced when David Kronfeld wrote articles and letters to veterinary journals that challenged the animal-safety studies conducted on the genetically engineered bovine growth hormone (rbGH). According to the dairy newspaper *The Milkweed*, "For his 'heresy', a Monsanto employee . . . wrote three letters to [Virginia Polytechnic Institute, the university where Kronfeld worked] during 1989 implicitly threatening that Monsanto might cease all research grants to that university if Kronfeld didn't silence his criticisms of bGH research."[40]

"Scientific experts cannot be expected to be independent and reliable advisors in safety issues considering the increasing dependence of science on financial support from the industry," writes Jaan Suurküla, M.D., in an editorial for PSRAST (Physicians and Scientists for Responsible Application of Science and Technology).[41] And a columnist in *New Scientist* warns, "Industry-based scientists have influence in high places— they move in the corridors of government."[14]

Industry-based scientists also appear to be well entrenched in the Rowett Institute, which, according to PSRAST, relies heavily on the profit of its commercial subsidiary, Rowett Research Services. This entity contracts with biotech, pharmaceutical and other companies for research contracts, the proceeds of which help fund the Institute. Thus, the Rowett is "dependent on the industry for its existence,"[42] and scientists like Arpad Pusztai depend on the Rowett for theirs.

In fact, scientists working for an institute usually cannot publish research without the institute's written permission prior to submittal. In order to get his work published in the *Lancet*, for example, Pusztai had to team up with a colleague at Aberdeen University who did further research on Pusztai's rat analysis. Only then could Pusztai "co-author" his study.

How many other scientists, like Arpad Pusztai, discovered unexpected problems with GM foods, but due to funding or employment considerations, chose not to pursue it? Why was Pusztai thrust into the spotlight?

Pusztai seems to have been propelled into the controversy due to his innocence and integrity. He was dedicated to thorough scientific inquiry and he thought everyone else was. He was a staunch believer in genetic engineering. When he discovered the damaging effects on his rats, Pusztai figured these problems could be worked out. He remained hopeful about the technology even after being suspended. As events unfolded, however, he began to realize how unscientific the business of science had become, when money, politics and reputation are at stake.

Pusztai says:

"In the last four or five years when I started to take these things seriously and I looked into similar cases, I became very concerned. The problems with GM foods may be irreversible and the true effects may only be seen well in the future.

"The situation is like the tobacco industry. They knew about it but they suppressed that information. They created misleading evidence that showed that the problem wasn't so serious. And all the time they knew how bad it was. Tobacco is bad enough. But genetic modification, if it is going to be problematic, if it is going to cause us real health problems, then tobacco will be nothing in comparison with this. The size of genetic modification and problems it may cause us are tremendous.

"If we injure the health prospects of humanity in this and the next and the next generation, then I think those people should be made accountable for the crimes they committed.

"Informing the public is the most important business in this very sorry affair, so one can do something."

Due to Pusztai's unexpected "popularity", he was approached by numerous scientists who quietly described their own surprise discoveries, further condemning the safety of GM foods. Some of these stories are described in this book. Others have to remain secret—for now.

Wisdom of the Geese

There's a farmer in Illinois who's been planting soybeans on his 50-acre field for years. Unfortunately, he also had a flock of soybean-eating geese that took up residence in a pond nearby.

Geese, being creatures of habit, returned to the same spot the next year to again feast on his soybeans. But this time, the geese ate only from a specific part of his field. There, as a result of their feasting, the beans grew only ankle high. The geese, it seemed, were boycotting the other part of the same field where the beans were able to grow waist-high. The reason: this year, the farmer had tried the new, genetically engineered soybeans. And you can see exactly where they were planted, for there is a line right down the middle of his field with the natural beans on one side and the genetically engineered beans, untouched by the geese, on the other.

Visiting that Illinois farm, veteran agricultural writer C.F. Marley said, "I've never seen anything like it. What's amazing is that the field with Roundup Ready [genetically engineered] beans had been planted to conventional beans the previous year, and the geese ate them. This year, they won't go near that field."[1]

Chapter 2

What Could Go Wrong?
A Partial List

In 1985 pigs were engineered with a human gene that produces human growth hormone. The scientists' goal was to produce a faster-growing pig. What they got was a freak show. "With their bristly hair and wide muzzles, these animals looked nothing like the pigs on the farm owned by my grandfather,"[1] wrote Bill Lambrecht, reporter for the *St. Louis Post Dispatch*. In one of the first litters born with the growth hormone genes, a female piglet had no anus or genitals. Some of the pigs were too lethargic to stand. Others had arthritis, ulcers, enlarged hearts, dermatitis, vision problems, or renal disease.[2]

This was an early example in a long line of experiments with unpredicted results. In fact, the single most common outcome of genetic engineering has been surprise.

- Scientists engineered tobacco to produce a particular acid. That's all they wanted: the acid and nothing else. But the plant also created a toxic compound not normally found in tobacco.[3]

- Monsanto engineered two types of cotton: one to withstand applications of their Roundup Ready herbicide; the other to produce its own pesticide called Bt. The plants were not supposed to have any other novel attributes. The first year that the GM cotton was planted, however, tens of thousands of acres malfunctioned. In Missouri, plants dropped their cotton bolls; others died on contact with the herbicide they were supposedly engineered to tolerate. In Texas, up to 50 percent of the Bt cotton failed to provide the predicted level of insecticide and "numerous farmers had problems with germination, uneven growth, lower yield and other problems."[4]

- Scientists who genetically modified yeast to increase its fermentation were shocked to discover that it also increased levels of a naturally occurring toxin by 40 to 200 times. In their paper, which was published in the *International Journal of Food Science and Technology*, the authors admitted that their results "may raise some questions regarding the safety and acceptability of genetically engineered food, and give some credence to the many consumers who are not yet prepared to accept food produced using gene engineering techniques."[5] They also pointed out that their yeast, which had not been inserted with foreign genes but rather with multiple copies of the yeast's own gene, was not "substantially equivalent" to normal yeast, as is assumed by many governments' GMO policies.[4]

- Oxford University scientists who were attempting to suppress an enzyme in a potato accidentally boosted its starch content. Professor Chris Leaver, department head for Plant Sciences said, "We were as surprised as anyone." Leaver noted, "Nothing in our current understanding of the metabolic pathways of plants would have suggested that our enzyme would have such a profound influence on starch production."[6]

Why is it that scientists who engineer organisms to create one effect more often than not end up with something altogether different? One reason is that there is a lot going on with gene expression that we don't understand. Another reason is that many of the main scientific principles that formed the basis for genetic engineering have since been proven false.

To understand the possible causes of the deformed pigs or toxic tobacco, or of Pusztai's toxic potatoes for that matter, we must understand the process of genetic engineering. And that explanation begins with DNA.

DNA

Deoxyribonucleic acid, or DNA, is found inside the nucleus of every cell. It is a complex—we're talking *really* complex—molecule with billions of atoms tightly wrapped in a double helix formation (picture a ladder twisted into a spiral). If uncoiled, a single DNA molecule would stretch almost ten feet. DNA has been compared to a super computer, a blueprint, and the central switchboard. In ways that are still largely a mystery, DNA tells the cell how to behave and carries information that is passed from generation to generation.

Just as computer software is based on a simple code of ones and zeros, DNA's software is made up of four recurring units called nucleotides or bases—arranged in pairs. And like software, it is the sequence of these units that carries the information.

Every cellular organism has DNA, although it differs in size, content and complexity. Human DNA has three billion base pairs.

In their efforts to crack this code, scientists have determined that in many higher organisms, only about 1 to 3 percent of the DNA molecule is made up of genes. A gene is a specific sequence of bases that function as a unit, carrying particular "orders" for the body (or mind). Our genes can determine the colour of our hair and eyes, the height of our body, and a myriad of other "traits".

Genetic Engineering is Not an Extension of Natural Breeding

The DNA of a species changes and evolves, in part, through sexual reproduction. Genes from the female and male are combined and interact in various ways so that some of each parent are expressed in the offspring.

DNA can also mutate. And in spite of very intelligent "fix it" molecules in the cells of many species whose job is to repair the DNA, some mutations will stick around and be passed on to the next generation.

For centuries, farmers, gardeners and livestock breeders have intentionally bred plants or animals in order to combine desirable traits. If one type of rice grows well, for example, and another is tastier, a breeder may cross the two in the hopes of creating tastier, hardier rice. Sometimes the offspring's DNA will fulfil the breeder's desires; at other times the traits just won't combine well—nature had other plans.

With genetic engineering, breeders have a whole new bag of tricks. Instead of relying on species to pass on genes through mating, biologists cut the gene out of one species' DNA, modify it, and then insert it directly into another species' DNA. And since virtually all organisms have DNA, scientists don't have to limit the source of their genes to members of the same species. They can search anywhere in the plant, animal, bacteria, even human world to find genes with desired traits, or even synthesize genes in the laboratory that don't exist in nature.

For example, a scientist knew of a species of Arctic flounder that was resistant to freezing in cold temperatures. He wanted his tomatoes to resist cold temperatures so they wouldn't die in frost. The scientist didn't have to wait for the unlikely event of the fish mating with the tomato. Instead, he figured out which gene in the fish keeps it from freezing and then inserted that gene into the tomato's DNA. The anti-freeze gene has never ever, ever existed before in a tomato. But now it's in the scientist's tomatoes and all their future offspring.

Biotech proponents regularly spin their technology as an extension of natural breeding. The US Speaker of the House, for example, said in March 2003, "since the dawn of time, farmers have been modifying plants to improve yields and create new varieties resistant to pests and diseases. . . . Biotechnology is merely the next stage of development in this age-old process."[7]

While it may be the new tool in the breeder's toolbox, many scientists are adamant that the technology is completely different, and must not be mistaken with traditional breeding practices. George Wald, Nobel Laureate in Medicine and former Higgins Professor of Biology at Harvard University, said that genetic engineering presents "our society with problems unprecedented not only in the history of science, but of life on the Earth. It places in human hands the capacity to redesign living organisms, the products of some three billion years of evolution. Such intervention must not be confused with previous intrusions upon the natural order of living organisms; animal and plant breeding, for example; or the artificial induction of mutations, as with X-rays. All such earlier procedures worked within single or closely related species. The nub of the new technology is to move genes back and forth, not only across species lines, but across any boundaries that now divide living organisms."

In Wald's view, the fact that a fish can't mate with a tomato is not random, but the result of the natural evolution of life on earth. By crossing that natural, age-old species barrier, genetic engineers are not simply changing a specific species. They are tampering with the evolution of all species. "The results will be essentially new organisms, self-perpetuating and hence permanent. Once created, they cannot be recalled."

Wald warned, "Up to now, living organisms have evolved very slowly, and new forms have had plenty of time to settle in. Now whole proteins will be transposed overnight into wholly new associations, with consequences no one can foretell, either for the host organism, or their neighbours."

Wald said that genetic engineering "presents probably the largest ethical problem that science has ever had to face." He also warned, "going ahead in this direction may be not only unwise, but dangerous. Potentially, it could breed new animal and plant diseases, new sources of cancer, novel epidemics."[8]

Genetic Engineering is Based on an Obsolete Model

When the scientist took the anti-freeze gene from the fish, he did so because he knew that the gene creates a particular anti-freeze protein. It's the protein that helps the fish to survive cold temperatures. Genes give their orders to the cell by creating proteins, which in turn confer the trait to the plant or animal.

The old theory of genetics asserted that each gene is coded for its own single unique protein. Biologists also estimated that the number of proteins in the human body was 100,000 or more. Thus, they predicted that there would conveniently be about 100,000 genes in human DNA. When the number of human genes was ultimately tallied and reported on June 26, 2000, it shocked the scientific world: there were only about 30,000. Oops.

This figure not only fails to account for the estimated number of proteins, it falls short of explaining the vast quantity of inheritable traits in the human body. Moreover, there are *weeds* with as many as 26,000 genes. Given the one protein-one gene theory, shouldn't humans have far more genes than a weed? Something seemed terribly wrong.

It turns out that the vast majority of genes do not encode for a unique protein. On the contrary, some genes can make many, many proteins. In fact, the current record is set by a single gene from a fruit fly, which can generate up to 38,016 different protein molecules.[2]

In humans, nearly all genes are theoretically able to make two or more proteins. The number of human genes capable of coding for only a single trait can be counted on your hands.

The fact that a gene creates multiple proteins may explain some of the surprises that keep popping up for genetic engineers, and it is first on our list of what can go wrong and why.

1. Code Scramblers

To make a protein, the DNA uses its unique genetic code to write a prescription for its chief assistant, RNA. The RNA fills the prescription by creating and assembling amino acids. The amino acids form the protein. But in some cases, before RNA fills the prescription for the protein, along come the spliceosomes (we'll call them code scramblers), a group of molecules that cut up the RNA, rearrange it and then reassemble it. Once reassembled (alternately spliced), the RNA now has an entirely new prescription resulting in the creation of an entirely new protein. The code scramblers can rearrange a single RNA code in many, many ways, "creating hundreds and even thousands of different proteins from a single gene".[2]

The code scramblers are by no means arbitrary in their work. Imagine roaming molecules, carefully observing passing RNA, comparing them to the pictures on a clipboard of their Ten Most Wanted. When a match is spotted, on jumps the scrambler, who quickly checks the pager on his belt—equipped with text messaging—to consult the latest list of *"proteins needed"*, . . . or something like that.

Now let's consider the anti-freeze gene making its debut in the DNA of a tomato. The gene writes a prescription for RNA, instructing it to make an anti-freeze protein. But what happens when that RNA wanders past a code scrambler? What if the scrambler checks its clipboard and thinks there's a match? If the scrambler jumps on the foreign RNA that it has never encountered before and starts to move things around, God knows what protein it will create. (Man certainly doesn't.)

As long as the scientists were absolutely sure that a single gene created one and only one possible protein, then they could confidently insert that gene in a new species and be sure that it would create that unique protein. The scientists were absolutely sure; but they were wrong.

According to Barry Commoner, senior scientist at the Center for the Biology of Natural Systems at Queens College, City University of New York, "The fact that one gene can give rise to multiple proteins . . . destroys the theoretical foundation of a multibillion-dollar industry, the genetic engineering of food crops." In the presence of the code scramblers, the foreign genes inserted in GM crops might create many unintended proteins "with unpredictable effects on ecosystems and human health."[2]

The relationship between genes and the code scramblers has evolved for billions of years, right along with the evolution of DNA itself. We do not fully understand how they work together in the same species. We certainly can't predict how they will work when a gene from one species meets a code scrambler from another. Will the code scramblers ignore the foreign gene? Or will the code scramblers try to switch around its prescription and accidentally create a protein that might be toxic, or allergenic, or the source of a new disease? It's hard to say; hard to say since no one generally tests for this.

"They don't want to know," says Joseph Cummins, professor emeritus of genetics at the University of Western Ontario. He says that in spite of the overwhelming evidence to the contrary, the biotech industry would rather make the assumption that *their* foreign gene will somehow avoid the host organism's scramblers. If not, genetic engineering would be way too risky.[9]

The engineers *might* be excused for not testing for new proteins when inserting genes taken from bacteria. Unlike genes from plants, animals and humans, bacterial genes are usually not scrambled. In order to be scrambled, genes need to be equipped with introns (we'll call them signal beacons). These beacons send out a message loud and clear to the code scramblers, saying in effect, "Pick me!" Most scientists assume that nearly all genes that have these signals end up getting scrambled, and that those that don't, do not. Most plant and animal genes have signal beacons. In bacteria, most do not.

Since bacterial genes don't often have the signals, scientists *assume* that they won't get scrambled when they are put into a different genetic environment. This should mean that genetically modified Bt crops are immune to scrambling. Bt crops, including maize, cotton and oilseed rape, are engineered to produce their own insecticide. The foreign gene that produces the Bt toxin is from a bacterium and is devoid of signal beacons.

But when engineers first put the Bt gene into plants, the gene didn't work very well; it produced very little Bt protein. To pump up Bt production, they attached—guess what—SIGNAL BEACONS! These signal beacons, it turns out, not only enable scrambling, they can also pump up protein production. Sure enough, the newly outfitted Bt genes did produce more Bt. The plants' genetics responded to the signals. But wouldn't that mean that the code scramblers would also respond?

Rather than doing a careful analysis to verify that unintended proteins were not created, the manufacturers of GM crops decided to stay with their original assumptions. They assume, according to Commoner, "without adequate experimental proof, that a bacterial gene for an insecticidal protein, for example, transferred to a corn [maize] plant, will produce precisely that protein and nothing else."[2]

Code scramblers aren't the only things found in a cell that can make an inserted foreign gene change its characteristics.

2. Hitchhikers

Even if the foreign gene gets past the code scramblers untouched and creates its intended protein, there's another problem. According to Professor David Schubert of The Salk Institute for Biological Studies, the effect that a particular protein has on a plant or animal "can be modified by the addition of molecules such as phosphate, sulphate, sugars, or lipids". These add-on molecules (call them hitchhikers) vary throughout the organism. "Each cell type expresses a unique repertoire"[10] of them, and may modify the protein in different ways. For example, the same protein found in both the liver and the brain can pick up entirely different hitchhikers and consequently have different effects on the body.

With Bt maize, will the foreign insecticide protein pick up a hitchhiker molecule in the maize kernel, changing the way it behaves? Will a different hitchhiker be picked up in the roots, or leaves, or stems, changing the protein's behaviour there? The answers are not known. Scientists don't necessarily know *if* the hitchhikers are added or *what* their effects on the plant might be.

3. Chaperones

In addition to its amino acid sequence and the presence of hitchhikers, a protein's shape also determines its effect. In order to do its job right, "the newly made protein, a strung-out ribbon of a molecule, must be folded up into a precisely organized . . . structure," says Commoner. He points out that according to the old theory of genetics, the protein "always folded itself up in the right way once its amino acid sequence had been determined. In the 1980s, however, it was discovered that some . . . proteins are, on their own, likely to become misfolded—and therefore remain biochemically inactive—unless they come in contact with a special type of 'chaperone' protein that properly folds them."[2]

Here again is a problem: What happens when a foreign insecticide protein comes face to face with the maize's chaperone folders. Will they leave it alone? Will they try to fold it? Will they get it right? There's no way to know. The chaperones have never met the protein before.

Dr. Peter Wills of Auckland University warns, "an incorrectly folded form of an ordinary cellular protein can under certain circumstances . . . [duplicate itself] and give rise to infectious neurological disease."[11] Prions, responsible for mad cow disease and the deadly Creutzfeld-Jacob disease in humans, are examples of such dangerous misfolded proteins.

So far we have identified three potential sources for unpredicted effects that were not taken into consideration by the crafters of genetic engineering: code scramblers, hitchhiker molecules and chaperone folders. These complex processes have, in Commoner's words, "evolved in a harmonious relationship over a long evolutionary period", subject to "many thousands of years of testing, in nature". But when you take a gene that is used to functioning in a bacterium and put it into the DNA of soya, cotton, or maize, for example, "the plant system's evolutionary history is very different from the bacterial gene's." What was a harmonious interdependence in their own environment "is likely to be disrupted in unspecified, imprecise and . . . unpredictable ways." In Commoner's view, "these disruptions are revealed by the numerous experimental failures that occur before a [GM] organism is actually produced and by unexpected genetic changes that occur even when the gene has been successfully transferred."

He concludes, "The biotechnology industry is based on science that is forty years old and conveniently devoid of more recent results." He says, "There are strong reasons to fear the potential consequences of transferring a DNA gene between species. What the public fears is not the experimental science but the fundamentally irrational decision to let it out of the laboratory into the real world before we truly understand it."[2]

Richard Strohman, professor emeritus at the University of California (UC), Berkeley, adds: "We're in a crisis position where we know the weaknesses of the genetic concept, but we don't know how to incorporate it into a more complete understanding. Monsanto knows this. DuPont knows this. Novartis knows this. They all know what I know. But they don't want to look at it because it's too complicated and it's going to cost too much to figure it out."[12]

4. Messing Up the Host's Normal DNA

We have used the word "insert" when describing the "placement" of foreign genes into a host DNA. That's more than polite. One common method used to "insert" genes is to blast them into the DNA with a 22-calibre gene gun. Scientists first coat thousands of tiny shards of gold or tungsten with the foreign gene. Then they point it at a dish containing thousands of unsuspecting cells. Then they fire, hoping that at least some of the foreign genes will end up in the right place in at least some of the

DNA. This, by the way, is what the biotech industry refers to as their highly precise method of gene transfer.

The impact of a gene-coated shard flying at hundreds of miles an hour into the DNA might, as you probably have guessed, result in some structural "consequences". The native genes can be damaged in ways that the engineer may not be able to identify.

When foreign genes take up residence in DNA, whether via gene guns or through other methods, it can have drastic effects. Michael Antoniou, senior lecturer in molecular pathology and head of a research group at one of London's leading teaching hospitals, says, "This procedure results in disruption of the genetic blueprint of the organism with totally unpredictable consequences."[11] The information in the DNA can be reorganized and mixed up.

"The phenomenon of rearrangements at the point of genetic insertion is widely recognized," admits Marcia Vincent, a Monsanto spokeswoman.[13] Her comment, however, understates the impact. The BBC's *Tomorrow's World Magazine* is more explicit: "Genetic engineering is generally a hit and miss affair. The genes may be inserted the wrong way round or multiple copies may be scattered throughout a plant's genome. They may be inserted inside other genes—destroying their activity or massively increasing it. More worryingly, a plant's genetic make-up may become unstable—again with unpredictable results. Genes may switch on or off unexpectedly with possible . . . unexpected or unknowable effects. Genes can hop around the genome for no obvious rhyme or reason. Rogue toxins may be produced or existing ones amplified massively. Such problems may only arise hundreds of generations after the crops are originally modified."[14]

DNA instability is a common feature of genetic engineering. In a survey of at least thirty companies developing GM crops, all had observed it.[15]

New DNA chip technology has recently allowed scientists to monitor changes in DNA functioning when foreign genes are inserted. In one experiment, there was a staggering 5 percent disruption of overall gene expression. In other words, after a single foreign gene had been added through genetic engineering, one out of every 20 genes that were creating proteins either increased or decreased their output. According to Schubert, "while these types of unpredicted changes in gene expression are very real, they have not received much attention outside the community of the DNA chip users." He adds that, "there is currently no way to predict the resultant changes in protein synthesis."[10]

A change in the host's DNA due to the process of inserting a foreign gene is called "insertion mutation". In human gene therapy, studies have verified that insertion mutation can lead to leukemia in children. Such an effect is so widely recognized, there is even a term, "insertion carcinogenesis", to describe it. In plants, according to Cummins, the disruptions may be similarly dangerous, producing unpredicted toxins.[9] But they haven't been studied closely.

5. Horizontal Gene Transfer and Antibiotic Resistance

After the foreign genes are blasted into the cells, only a small percentage end up inside DNA. To figure out which of the thousands of cells on the plate have the foreign gene in their DNA, scientists typically attach an Antibiotic Resistant Marker (ARM) gene to their foreign gene. If this gene package makes it into the DNA, the ARM gene will render that cell invincible to a normally deadly dose of antibiotics.

Thus, after the genes are shot into the pile of cells, the cells are all doused with antibiotics. Those that survive got the genes in their DNA. Those that die did not. Only one in thousands survives.

Many scientists are concerned that when humans and animals eat GM food, the ARM genes will transfer into the bacteria found inside the digestive system. This process, whereby genes travel from one species to another, is called "horizontal gene transfer". If the ARM gene moves between species it could result in new and dangerous antibiotic-resistant diseases. The British Medical Association mentioned this serious risk as one of the reasons why they called for an immediate moratorium on genetically engineered foods.

The biotech companies assure the public that ARM genes cannot be transferred between food and bacteria in the human gut. They refer to evidence, says Michael Hansen, from animal studies in the 1970s and '80s that "failed to find evidence that DNA survived digestion."[4] When detection techniques became more sensitive starting in the late 1980s, however, animal feeding studies confirmed that DNA not only survives, it is found in the blood, intestinal wall, liver, spleen and faeces, and even remains intact in the digestive system for more than five days. DNA can even travel via the placenta into unborn mice. More pertinent, however, is a 2002 study that was dubbed "the world's first known trial of GM foods on human volunteers". Research demonstrated that "a relatively large proportion of genetically modified DNA survived the passage through"[16] the human small intestine and that horizontal gene transfer did occur. GM soya that was present in the burger and milkshake fed to subjects at the beginning of the experiment transferred its herbicide-resistant gene to the bacteria inside their digestive systems. The transfer occurred after only a single meal.

"Everyone used to deny that this was possible," says Antoniou. "It suggests that you can get antibiotic marker genes spreading around the stomach which would compromise antibiotic resistance. They have shown that this can happen even at very low levels after just one meal."[16]

Bt maize contains an ARM gene that resists the commonly prescribed antibiotic, ampicillin. Scientists worry that this gene's widespread presence in human and animal food will render ampicillin useless in treating disease. The World Health Organization, Britain's House of Lords, the American Medical Association, and even the Royal Society have all called for a phase-out of the use of ARM genes.

6. Position Effects

When a foreign gene makes it into the DNA, there is no telling where along the strand it will end up. The inserted gene could disrupt any number of naturally expressed traits depending on where it lands. For example, when scientists inserted a foreign gene into a plant from the mustard family, the plant's ability to crossbreed with related species varied depending on where in the DNA the gene was located.[17] Similarly, the location of a foreign gene can dictate how well it does its job. In some locations it will not produce its protein at all; in others, it will produce too little. These location-specific changes are called "position effect"—a kind of genetic Russian roulette.

7. Gene Silencing

One common position effect is that either the foreign gene or the native genes in their vicinity get shut off; they are no longer able to produce their protein. This common and unpredictable occurrence is called "gene silencing".

One way that a native gene can get permanently disabled is if the foreign gene ends up right in the middle of it. This happened in one experiment and the mouse embryos ended up dying.[4]

Silencing native genes can result in all sorts of unpredictable outcomes. For example, in his testimony before the United States Environmental

Protection Agency (EPA), Michael Hansen of the Consumers Union warned that if the process of genetic engineering "turned off" a native gene whose job was to prevent "the expression of some toxin, the net result of the insertion would be to increase the level of that toxin."[4]

8. Environmental Influences

Scientists observed gene silencing when genetically engineering petunia plants. The inserted foreign gene was designed to express salmon red. Scientists expected virtually all the flowers to bloom with the same red colour. Instead, the flowers varied in both colour and pattern. The variation was due to the silencing of the foreign genes in some of the plants. Which plants had silenced foreign genes depended on the position effect—where in the DNA those foreign genes ended up.[18]

In this experiment, however, there was another factor influencing the plants. The colour of those petunia flowers inexplicably changed during the season. More of the foreign genes were switched off as the season progressed. Here, the changes in gene expression were apparently linked to environmental changes.

9. Light Switches—Turning On Your Genes (at Random?)

In normal circumstances, a gene in one cell will busily pump out its protein, while in another cell, that same gene just quietly hangs out, unused; its protein isn't needed. Take, for example, the gene whose protein makes the eyes blue. In the pigment cells of the iris, that gene stays busy. But in the whites of the eyes, that same gene gets to relax. Otherwise, if it got activated, perhaps the entire eye would turn blue.

Who tells the gene when to work and when to rest? Somehow every cell provides a clear-cut job description for all its genes. Work here; rest there; work for a little, then take a break. And the job description can change depending on what the body needs.

When genetic engineers put an insecticide gene into the DNA of maize, however, the maize cell doesn't have a clue what to do with this gene that it's never seen before. Should it be turned on or turned off? Biologists can't speak the language of the cell. They don't know how to tell it to monitor the whole organism and to switch on the new gene only when needed—as it does with all the other genes. Instead, biologists do something unprecedented in the cell's experience. The new gene is sent in with a "light switch" permanently in the "on" position, set to high intensity. This keeps the new gene working 24/7, non-stop, in all cells of the plant. The light switch, called the "promoter", consists of genetic material that is attached to the insecticide gene prior to insertion.

The selection of this genetic material presents an interesting *and dangerous* challenge. The cell protects DNA from foreign invaders. In plants and animals, an elaborate defence system normally prevents foreign genes from getting a foothold. But there are certain highly aggressive genetic invaders that get past the cell's defences. Most notable among these are viruses, some of which are cancer-causing. These can wreak havoc on the DNA and the entire organism.

Molecular biologists borrow the light switch from one of these viruses, since it works in the DNA of all types of plants. Called the Cauliflower Mosaic Virus (CaMV) promoter, "it is designed to overcome a plant cell's defensive devices to prevent foreign DNA from being expressed," says Hansen. The CaMV's light switch or promoter is a key element enabling the virus to "hijack a plant cell's genetic machinery and make many copies of itself".[4]

This bullying nature allows it to operate independently of the cell's normal, harmonious and coordinated self-regulation. Therefore, in spite of any protests by the cell or its DNA, the CaMV promoter will cause the gene to which it is attached to switch into overdrive.

Some biologists warn that the energy and resources that a plant requires to keep a gene switched on round the clock in every cell can drain other systems. There's no way to know which other systems will get sacrificed or what consequences that will mean for the health of the plant (or for the eater of the plant).

But the potential danger of the CaMV promoter is far greater. Hansen told an EPA panel that since this promoter operates "outside of normal regulatory circuits" of the plant's own DNA, it "may be one of the reasons why [GM foods] are known to be so unstable."[4] In fact, Dr. Pusztai's team suspected that it is the promoter's unstable, unregulated and aggressive nature that caused the immune and organ damage of his rats. Scientists from all over the world have expressed concern about CaMV, calling for an immediate ban.

Their concerns have been heightened by studies showing that the CaMV promoter not only turns on the foreign gene to which it is attached, but other native genes as well. In other words, genes that are supposed to be dormant, like the blue-eye genes in the whites of the eyes, are forced to start producing their proteins against the wishes of the cell. The CaMV promoter may turn on native genes "over long distances"[4] up and down the strand of DNA. It can even turn genes on in a different chromosome (section of DNA). There's no practical way to turn off or adjust the volume of these chemical switches.[17] This can create a flood of proteins that are totally inappropriate.

Turning genes on or off is another form of Russian roulette. Whether the process creates new toxins, allergens, cancers, or nutritional changes is anyone's guess.

10. Hot Spots

Studies also show that the promoter creates a "hotspot" in the DNA. This means that the whole DNA section, or chromosome, can become unstable. This can cause breaks in the strand or exchanges of genes with other chromosomes. According to Cummins, a promoter can have "the same impact as a heavy dose of gamma radiation".

11. Waking Sleeping Viruses

The nature of the CaMV promoter presents yet another risk, which Cummins believes is "probably the greatest threat from genetically altered crops". He says that laboratory research demonstrates that "the insertion of modified virus and insect virus genes into crops" can "create highly virulent new viruses."

To understand this, we must again look at how the theory of genetics has evolved since genetic engineering began. Only a small percentage of the DNA has been identified as genes. In humans, it's between 1.1 and 1.4 percent. The much larger portion of the DNA was once referred to as "junk DNA". It was considered by scientists to be useless debris left over from the evolution of the species. Shooting foreign genes into sections of junk DNA was considered a safe zone. In reality, it may be just the opposite.

As the DNA has evolved, it has become a repository of genetic material going back eons. Included among this material are viruses that have worked their way into the DNA in the distant past, but are now dormant. "Most viruses have eroded," says Cummins, "and have lost the ability to become reactivated as viruses." But he warns, "some are quite complete and would be easy to turn on."[9]

Cummins and others are concerned that the CaMV promoter, which is used in nearly all commercialized GM crops, might be reactivating viruses. In addition to waking viruses in the DNA of maize, soya and other GM foods, they are concerned that the promoters might move between organisms through horizontal gene transfer. Suppose, for example, that the CaMV promoter from a GM maize kernel wanders off inside the stomach of a human and gets reattached to the DNA of a dormant virus. Instead of promoting an insecticide gene as was intended, it may now be switching on a virus.

In their paper, "Cauliflower Mosaic Viral Promoter—A Recipe for Disaster", Ho, Ryan and Cummins warn, "Horizontal transfer of the CaMV promoter . . . has the potential to reactivate dormant viruses or [create] new viruses in all species to which it is transferred."[19]

12. Cancer

The CaMV light switch and other viral promoters used in GM crops can also activate other, non-viral genes in the species where it "happens to be transferred," says Ho and others. "One consequence of such inappropriate over-expression of genes may be cancer."[19]

Stanley Ewen, one of Scotland's leading experts in tissue diseases, agrees. He says the CaMV promoter "could affect stomach and colonic lining by causing a growth factor effect with the unproven possibility of hastening cancer formation in those organs." Ewen, who had collaborated with Pusztai on the *Lancet* publication, may have seen first-hand early signs of such growth in the rats' thickened intestines. In fact, all three studies that reported unusual cell proliferation (described in the last chapter) might have been identifying the effects of the CaMV promoter.

In December 2002, Ewen issued a strong warning to the Scottish Parliament's Health and Community Care Committee, which was considering the fate of future trial plots of GM crops. Ewen said that even the food and water in the area near the crops may be contaminated by GM material. He also described risks of GM animal feed. "It is possible cows' milk will contain GM derivatives that can be directly ingested by humans as milk or cheese. Even a lightly cooked, thick fillet steak could contain active GM material."

Thorough cooking would probably destroy most GM material. Stomach acids might similarly break it down (although evidence presented in a later chapter disputes this). Ewen is concerned that those who have impaired digestion as a result of even common stomach infections might be more at risk from intact GM genes and would be vulnerable to the

CaMV promoter's growth factor effect.

"I don't want to be scare-mongering, I want to be understated," Ewen said. "But I'm very concerned that people who rely on local produce might be endangering themselves."[20]

13. Risks from Breathing Genetically Modified DNA

In the summer of 2003, at least thirty-nine people living adjacent to a Bt cornfield in the Philippines developed respiratory, intestinal, and skin reactions while the maize was pollinating. Blood tests conducted by the Norwegian Institute for Gene Ecology verified antibody reactions to Bt toxin, indicating an immune reaction to the pollen. Results are preliminary and it isn't certain if the symptoms are related to the maize. Years earlier, the UK Government's Joint Food Safety and Standards Group wrote to the US FDA about the potential dangers of inhaled GM pollen, even warning that genes might transfer into humans.

While the above study looked at reactions to Bt, not gene transfer, other research by the Norwegian Institute discovered intact CaMV promoters inside rat tissues two hours, six hours, and three days after genetic material was mixed into a single meal. They also verified that the CaMV promoter functions in human, rat, and fish cells, inside test tubes. These findings overturn industry assertions that horizontal gene transfer can't happen and that the promoter only functions in plants.

Although the CaMV is found in cauliflower and other vegetables, according to Mae-Wan Ho, a geneticist and biophysicist formerly at the Open University, the viruses found naturally in vegetables are protected by a protein coat that is wrapped around the DNA. This prevents the CaMV from entering the cells of mammals. The CaMV promoter in GM foods, however, is naked viral DNA, with no such restrictions.[21]

Other Unknowns

Genetic engineering is built on a long list of assumptions. The main assumption is that foreign genes will always operate the same way in the new host organism. Here are four additional challenges to these assumptions.

14. Synthetic Genes:
Most foreign genes used in GM crops are not natural. They are synthetic. Since plant and bacteria genes use different sequences to "describe" certain amino acids, the codes of bacterial genes

have to be altered so they will "read" correctly in the plant. Cummins says, "Use of synthetic genes has become pervasive in genetic engineering and the synthetic genes are assumed to be equivalent. But there are a lot of differences between them that have been ignored. In particular, the bacterial genes used in Bt crops and Roundup Ready soya and maize are changed a great deal." He says governmental agencies simply accept the companies' assumptions of equivalence since "the regulators are naive in the area of genetics and molecular biology."[9]

15. Genetic Disposition: For reasons not well understood, inserting the same gene into different varieties of the same plant species can have widely varying results. According to his testimony to the EPA Science Advisory Panel in October 2000, Hansen says, "In some varieties, the trait can be expressed at high enough levels to have the desired impact. In others, the expression level is too low to have the desired impact."[4] Similarly, varieties of the same species may be prone to dangerous side effects when a new gene is inserted. The unpredictable influence of genetic disposition is not usually addressed in safety studies.

16. Complex Unpredictable Interactions: "When you insert a foreign gene, you are changing the whole metabolic process," says the University of Georgia's Sharad Phatak. "You just don't change one thing. Each change is going to have an effect on other pathways. Will any one gene kick off a whole slew of changes? We don't know for sure."[22]

Genes can influence each other. Proteins can influence each other. And altered proteins can activate or de-activate genes. With each change, a new interaction can begin, setting off yet more changes. This type of unpredicted chain reaction may have produced the toxin responsible for the deadly epidemic described in the next chapter.

17. Rearranged codes: Sometimes the process of genetic engineering results in a rearranged sequence of genetic information. Although the cause is not clear, it may be associated with the effect of the gene gun combined with the cell's attempt to repair wounds.

18. Gene Stacking: The opportunity for unpredicted interactions increases manifold when GM crops are engineered with not just one foreign gene, but with multiple "stacked" genes. One version of Monsanto's New Leaf potato, for example, was stacked with eight different traits—it created its own pesticide, resisted diseases, was tolerant to herbicide, increased its weight and reduced bruising.[1] Some GM crops *accidentally* acquire additional foreign genes through cross-pollination. Oilseed rape plants in Canada, for example, ended up with foreign genes from two different companies, each conferring tolerance to its own brand of herbicide.

Stacked genes and their proteins may interact in dangerous ways. Traditional pesticides illustrate this principle. When they are mixed with other pesticides or chemicals, their strength can be multiplied. "Compounds that enhance the activity of pesticides are not uncommon." In fact, scientists accidentally discovered that the Bt toxin created by varieties of GM maize, cotton and oilseed rape becomes "more deadly" to insects when mixed with very small amounts of a naturally occurring antibiotic—a byproduct of bacteria. Tests have not been conducted to determine if the "greatly enhanced"[23] toxicity is dangerous to humans or wildlife.

19. Nutritional Problems

Changes in the DNA—both intended and accidental—can influence a plant's nutritional content. In fact, many of the potential problems already addressed in this list might change the health value of a GM food. Studies have pointed out numerous differences in the composition of GM maize and soya compared to their natural non-GM counterparts. Altered nutrition can lead to unanticipated side effects. Cows fed GM Roundup Ready soya, for example, produced milk with increased fat content.[24] This illustrates a cascading effect, where one problem leads to others.

20. Allergens

Genetic engineering can transform a harmless food into one containing a potentially deadly allergen in at least three different ways: 1. The level of a naturally occurring allergen might be increased; 2. A gene taken from one type of food might transfer allergenic properties when inserted into another food; and 3. Unknown allergens may result from foreign genes and proteins never before part of the human food supply. This serious topic is discussed in greater detail in its own chapter.

21. Human Error

In addition to working with obsolete theories and limited understanding, genetic engineers also operate in a field where there are ample opportunities for human error. Some errors are caught. Some get away.

One that got away was reported on February 21, 1999. According to the *Independent on Sunday*, Monsanto had mixed up "crucial information" about a foreign gene that was inserted into herbicide-tolerant maize. The incorrect data had been submitted to the UK's Advisory Committee on Releases into the Environment (ACRE) for a safety assessment. Committee "members were furious that Monsanto had asked them to approve a marketing application based on inaccurate information," reported the article. "They accused Monsanto of submitting sloppy research, 'poor interpretation' and work far below required standards." Monsanto was referred to as "incompetent" and their standard of work "wholly unacceptable".

"It's very worrying," said Janey White, a molecular biologist. "This means that somebody somewhere in Monsanto is getting it wrong." Apparently, the mistake had already got past regulators in the United States, where the corn [maize] was already approved.[25]

In addition to errors in the creation or evaluation of a GMO, another type of common error is accidentally letting unapproved GMOs into the food supply. For example, in February 2003, Reuters reported, "Nearly 400 pigs used in US bioengineering research may have entered the food supply because they were sold to a livestock dealer instead of being destroyed."[26] Similarly, a year earlier eleven GM piglets had been accidentally ground into poultry feed. There have also been numerous incidences of unapproved crop varieties ending up in food: StarLink maize was the most famous example. GM crops modified to produce pharmaceuticals or industrial chemicals have also contaminated nearby fields through pollen transfer or accidental mixing.

Case Study: Roundup Ready Soybeans

A combination of human error and the unpredictable effects of genetic engineering was revealed in May 2000. Monsanto's Roundup Ready soybeans had already been on the market for seven years. The company *thought* they had inserted only a single foreign gene (along with its CaMV promoter). The gene, derived from bacteria, allowed the soya plant to survive high concentrations of Monsanto's herbicide called Roundup. To the company's surprise, they discovered that there were two additional gene fragments that had been inserted into the soya DNA accidentally.

According to Sue Mayer, Director of the independent research group Genewatch, "These results demonstrate that genetic modification is a clumsy process, not precise as is often claimed. There is no control over how many genes, in what order, or where they are inserted."[27]

She added, "Additional copies or fragments of genes may affect the operation of the other inserted genes, which could have consequences for the performance and composition of the plant. This may have implications for human and environmental safety."

Charlie Kronick of Greenpeace added, "After years on the market, Monsanto reveals that neither the industry or the regulators actually know what genes are in it. What else don't we know?"

More that we didn't know was soon revealed. A year later, a team of Belgian scientists published their surprising discovery that adjacent to one of those rogue inserted gene fragments was a sequence of DNA—534 bases—that was not part of the Roundup gene and was not natural soybean DNA either.[28]

According to the *New York Times*, their findings "suggested that this unknown DNA is probably the plant's own DNA but that it was somehow rearranged, or scrambled, at the time the bacterial gene was inserted. Another possibility, they said, is that a portion of the plant's DNA was deleted, leaving other DNA in that position."[29]

Commoner cites a third possibility: The plant's own proteins, which are normally used to correct DNA errors, might have rearranged the foreign gene's sequence of bases.[2]

Whatever the reason, "The abnormal DNA was large enough to produce a new protein, a potentially harmful protein."[2]

Doug Parr, Greenpeace-UK's chief scientific adviser, warned, "No one knows what this extra gene sequence is, what it will produce in the soybean, and what its effects will be."[30]

Tony Combes of Monsanto defended the newly discovered piece of DNA, saying, "It would have been a constituent of the Roundup Ready soybeans used in all the safety assessment studies."

Safety Assessment?

What safety assessment studies? According to Arpad Pusztai, he hasn't yet seen studies that were adequate to identify potential hazards from even the *intended* foreign gene, let alone gene fragments or re-scrambled DNA.

Let's be specific. Let's look at the body of safety assessment research on these soybeans, including both published studies and the unpublished research Monsanto submitted to the UK's ACNFP for approval. We'll see if it adequately tests for the potential risks discussed thus far in this chapter.

1. To make sure that **code scramblers** didn't rearrange the code on Monsanto's foreign genes and create new, unexpected proteins, researchers would have been required to identify the type and quantity of all proteins in the soybean, both before and after modification. This pre- and post-inspection would also be required to make sure that the **CaMV** light switch didn't accidentally turn on any native genes in the soy's DNA. The researchers did not do these tests.

2. To protect against the unintended behaviour of **hitchhiker molecules**, researchers would have been required to make sure that their new protein was devoid of these added molecules. And they would need to look for them in every part of the plant and in multiple growing conditions. They did not.

3. To avoid a misfolding of the new protein by the soy's **chaperone folders**, scientists would be required to compare the shape of their protein in the soya plant with its natural shape in the bacterium, also under varied conditions. They didn't do that either.

4. Scientists would have to carefully inspect the GM plant's entire DNA structure to be sure that the **process of inserting the gene** or the **CaMV hot spot** didn't disrupt any other sequence. They obviously didn't do that since they had missed two fragments of foreign DNA and the mystery DNA sequence that no one had ever seen before.

5. The fact that the **position effects** of the foreign gene and other factors can cause **gene silencing**—accidentally turning off native genes—presents a particularly difficult challenge. Some native genes are only expressed under very limited circumstances or in small regions of the plant. If one of these rarely used genes got silenced, how would researchers know? They would have to compare protein expression of all parts of the natural and GM plants under an enormous variety of circumstances, varying age, disease, nutrients, environment and pests, to name a few. Such a thorough analysis might not even be possible. Needless to say, it wasn't attempted.

6. Researchers also failed to safeguard against the creation of **new viruses**, which, in theory, might occur by either activating the host's sleeping viruses or through **horizontal gene transfer**.

7. **Antibiotic resistance** is not an issue with Roundup Ready soya. It is an issue for GM corn [maize] and it has not been adequately tested.

8. **Effects of the environment** on gene expression and **differences in genetic make-up** of soya varieties were assessed only under limited

conditions. Their effects were measured using only a few factors, such as crop yield.

9. **Synthetic genes** were considered equivalent. Any potential differences were not addressed in the research.

10. Very little research has evaluated unanticipated changes due to **complex interactions** or to multiple foreign genes, either produced by **gene stacking** or through **cross-pollination**.

11. Researchers did not look for **transfer of genetic material** via ingested **meat** or **milk**, through contaminated **water**, or by **inhaling GM pollen**. They assume none of these transfers are possible.

12. Scientist's concerns that the **CaMV promoter** might promote cell growth and lead to **cancer** have not been ruled out through rigorous studies.

13. Although some **nutritional** studies have been carried out, none have been exhaustive enough to identify the numerous differences that may be present.

14. Finally, researchers failed to adequately test to see if anyone would be **allergic** to their soya. Although some cursory analysis of its potential allergenic properties was done, a later chapter will reveal that no adequate test has yet been devised. This lack of safeguards has prompted Pusztai to label allergies as the "Achilles heel of GM food".

Thus, whatever the safety assessment the Monsanto representative was referring to, none has adequately identified or prevented many of the potentially serious problems that might already be plaguing society.

In our pursuit to discover the cause of hairy pigs, toxic tobacco and scores of other mishaps, we turned up more questions than answers. We can better understand the conclusions of a team of scientists who set out to document all that was not yet understood in the science of genetically modified crops. They said: "Controversies and knowledge gaps appear to be present at all levels."[31]

These gaps in knowledge are not merely academic. "Given our current lack of understanding of the consequences of [GM] technology," Schubert says, "GM food is not a safe option."[10] Commoner warns, "None of [the] essential tests are being performed, and billions of transgenic plants are now being grown with only the most rudimentary knowledge about [their changes]." He says, "Given that some unexpected effects may develop very slowly, crop plants should be monitored in successive generations as well."

"Without detailed, ongoing analyses," Commoner continues, "there is no way of knowing if hazardous consequences might arise. . . . The genetically engineered crops now being grown represent a massive uncontrolled experiment whose outcome is inherently unpredictable. The results could be catastrophic."[2]

Among the catastrophes that can occur is the creation of new toxins. "The unexpected production of toxic substances has now been observed in genetically engineered bacteria, yeast, plants and animals with the problem remaining undetected until a major health hazard has arisen," says Antoniou. "Moreover, [GM food and GM food processing agents] may produce an immediate effect or it could take years for full toxicity to come to light."[32]

In a later chapter, we look at one such deadly health hazard that might have taken many more years to discover, if not for the unique and acute symptoms of the disease, the unusual detective work of an alert physician, and lots of luck.

Wisdom of the Cows

In 1998, Howard Vlieger harvested both natural maize and a genetically modified Bt variety on his farm in Maurice, Iowa. Curious about how his cows would react to the pesticide-producing Bt maize, he filled one side of his sixteen-foot trough with the Bt and dumped natural maize on the other side. Normally, his cows would eat as much maize as was available, never leaving leftovers. But when he let twenty-five of them into the pen, they all congregated on the side of the trough with the natural maize. When it was gone, they nibbled a bit on the Bt, but quickly changed their minds and walked away.

A couple of years later, Vlieger joined a room full of farmers in Ames, Iowa to hear presidential candidate Al Gore. Troubled by Gore's unquestioning acceptance of GM foods, Vlieger asked Gore to support a recently introduced bill in congress requiring that GM foods be labelled. Gore replied that scientists said there is no difference between GM and non-GM foods. Vlieger said he respectfully disagreed and described how his cows refused to eat the GM maize. He added, "My cows are smarter than those scientists were." The room erupted in applause. Gore asked if any other farmers noticed a difference in the way their animals responded to GM food. About twelve to fifteen hands went up.[1]

"If a field contained GM and non-GM maize, cattle would always eat the non-GM first."[2] —Gale Lush, Nebraska

"A neighbour had been growing Pioneer Bt corn [maize]. When the cattle were turned out onto the stalks they just wouldn't eat them."[2]
—Gary Smith, Montana

"While my cows show a preference for open-pollinated corn [maize] over the hybrid varieties, they both beat Bt-corn hands down."[2]
—Tim Eisenbeis, South Dakota

According to a 1999 *Acres USA* article, cattle even broke through a fence and walked through a field of Roundup Ready maize to get to a non-GM variety that they ate. The cows left the GM maize untouched.[3]

Chapter 3

Spilled Milk

"The scientists' testimony before a [Canadian] Senate committee was like a scene from the conspiratorial television show 'The X-Files.'"[1] This was how Canada's leading paper, the *Globe and Mail*, described the story of six Canadian government scientists who tried to stand up to pressure to approve a product they believed was unsafe. The six were employed by Health Canada—the Canadian equivalent of the FDA. In 1998, they reviewed the recombinant (genetically engineered) bovine growth hormone (rbGH), which, when injected into dairy cows, increases milk production by 10 to 15 percent. Their job was to determine if it was safe for people to drink the milk from treated cows. They didn't think so, but senior Canadian officials and the product's maker Monsanto, tried to force them to approve it anyway.

The *Ottawa Citizen* reported the scene this way: "The senators sat dumbfounded as Dr. Margaret Haydon told of being in a meeting when officials from Monsanto, Inc., the drug's manufacturer, made an offer of between $1 million and $2 million to the scientists from Health Canada—an offer that she told the senators could only have been interpreted as a bribe."

The senators listened as Haydon "recounted how notes and files critical of scientific data provided by Monsanto were stolen from a locked filing cabinet in her office."[2] And, with her voice quavering, Haydon said that when she refused to approve the drug due to her concerns for human health, she was taken off the case.

The scientists told the Senate committee, "pharmaceutical manufacturers have far too much influence in the drug approval process." Scientists "often feel that their careers are threatened if they stand in the way of a drug they don't believe is safe." And "managers without scientific experience regularly overrule their decisions."[1]

One of the whistle-blowing scientists to testify, Shiv Chopra, revealed that the policy in the department is to "serve the client". The client, however, is no longer defined as the public: "The client is now the industry."

"We have been pressured and coerced to pass drugs of questionable safety, including [rbGH],"[2] Chopra reported to the committee. He "testified that one of his managers threatened to ship him and his colleagues to other departments where they would 'never be heard of again' if they didn't hurry favourable evaluations of rbGH."[3] He added that all files relating to rbGH were "now controlled by one senior bureaucrat and can only be viewed by gaining permission."[2] This was unique to rbGH; no other files had such limited access.

Senator Eugene Whelan responded, "I can't even believe I'm in Canada when I hear that your files have been stolen and that all the files are now in the hands of one person. . . . What the hell kind of a system have we got here?"[2]

The kind of system was further revealed when, after testifying, Chopra returned to his duties and was soon suspended for five days without pay. The cause for suspension, he later told another Senate committee, was retaliation for his testimony.

What was happening to the Canadian scientists in 1998 amounted to "re-runs" of what US government scientists faced in the 1980s, when the FDA was evaluating rbGH. A look at this drug's approval process gives us an example of the conflicts of interest, sloppy science and industry influence that also plague GM food policy. It's important to note, however, that the FDA treats GM food entirely differently from drugs. While the FDA spent several years evaluating rbGH, there is virtually no safety testing required for GM foods.

Monsanto's rbGH was officially approved for commercial release in February 1994. But the FDA had already declared it safe for humans back in 1985, which had allowed Monsanto to sell milk and meat from its research herds and experimental dairies. According to a document that was stolen from the FDA and later published in *The Milkweed*, they based their initial approval on a single twenty-eight-day rat feeding study along with some faulty assumptions about the "characteristics and biological activity" of the hormone. Not everyone at the agency was convinced the milk was safe. A few spoke out.

FDA chemist Joseph Settepani was in charge of quality control for the approval process of veterinary drugs. At a public hearing held by a New York congressman, Settepani described "a systematic human food-safety breakdown at the Center for Veterinary Medicine (CVM)". Soon after he spoke out, "he was reprimanded for insubordination, threatened with dismissal and stripped of his duties as a supervisor." In later testimony before a congressional subcommittee, Settepani said, "I was sent to a trailer at an experimental farm . . . completely isolated from agency policy-making on human food safety." He charged, "Dissent [at CVM] is not tolerated if it could seriously threaten industry profits."[4]

A second FDA scientist, Alexander Apostolou, came up against a similar response. Apostolou had become director of the Division of Toxicology after a distinguished career in industry and academia. In an affidavit he said, "Sound scientific procedures for evaluating human food safety of veterinary drugs have been disregarded. I have faced continuous pressure from my CVM superiors to reach scientific conclusions favourable to the drug industry. . . . In my time at CVM I have witnessed drug manufacturer sponsors improperly influence the agency's scientific analysis, decision-making and fundamental mission." Apostolou also described "the agency's clear trend to keep the industry content through uncritical acceptance of sponsor's claims and data, and in bending the rules to make their data look acceptable."[4]

When Apostolou began expressing his concerns at the agency, the reaction was hostile. "They assigned him to an impossible task, then put him on notice of poor performance when he allegedly failed to accomplish it—a common technique for neutralizing whistle-blowers."[4] Apostolou left the agency.

Apostolou said that FDA reviewers were more persuaded by the "quantity of reviews over the quality of the testing". Industry knew this. When four companies each sought approval of their own versions of rbGH, they buried the FDA in a mountain of information. According to Monsanto, their submission alone amounted to a stack of papers sixty-seven feet high.

"We were overwhelmed by the magnitude of the research," said veterinarian Richard Burroughs, who had a lead role in the review process of rbGH. At one point, the Human Safety Division reviewed forty volumes of submissions in just two weeks.

Burroughs acknowledges that the science in the studies was well outside the expertise of FDA employees. Rather than admit that they were in over their heads, Burroughs says, "the Center decided to cover up inappropriate studies and decisions." Officials "suppressed and manipulated data to cover up their own ignorance and incompetence."

As with other drug approvals, the FDA did not carry out its own tests on rbGH. Rather, the biotech industry performed the tests, summarized the data and presented it to FDA reviewers. Burroughs wrote the original protocols for evaluating the safety of rbGH on cows. As part of the process, he says, the drug developer "would come in and try to negotiate the protocols to water them down. . . . Then, if things didn't work out in their tests the way they wanted them to, they'd come in, present their data and see what they could salvage that would eventually help them market their product."[4]

But not all the data made it into the hands of the FDA. According to Burroughs, cows that developed infections of the udder, for example, were often dropped from the study, skewing the studies' conclusions. While the FDA and Monsanto said that they saw only a handful of cows develop udder infections, documents obtained through the Freedom of Information Act revealed that 9,500 cows from 500 farms were infected. In addition to omissions of data, dubious statistics were applied that further masked the drug's effects.

Burroughs, who was the only one in the unit who had real dairy herd experience, was involved with the rbGH approval process for nearly five years. When he saw what he considered to be unacceptable compromises on safety, he made his opinions known. When, for example, he learned that cows were going to be tested for only a single milking cycle, Burroughs insisted that two years was the minimum. He needed to see if the drug had an effect on newborn calves and on subsequent periods of lactation. Burroughs' requirements were frustrating the industry, which was anxious to get their products to market. When he called for toxicology and immunology tests in late 1989, that apparently proved to be too much. About a month later he was fired. He told the Humane Farming Association, "I was told that I was slowing down the approval process."[5] After he left the agency, they cancelled the toxicology study he had requested.

The retaliations against whistle-blowers like Settepani, Apostolou and Burroughs, did not go unnoticed among other employees. Thus, when others at the agency wanted to expose what was happening, they resorted to an anonymous letter to members of Congress. They wrote, "We are afraid to speak openly about the situation because of retribution from our director, Dr. Robert Livingston. Dr. Livingston openly harasses anyone who states an opinion in opposition to his." The letter was written on March

16, 1994, in response to the FDA decision not to label milk that comes from rbGH-treated cows.

[In the following excerpt of their letter, the employees refer to rbGH as BST. This stands for bovine somatotropin, the name that Monsanto adopted in the late 1980s after realizing that the word "hormone" was controversial.]

> "The basis of our concern is that Dr. Margaret Miller, Dr. Livingston's assistant and, from all indications, extremely 'close friend,' wrote the FDA's opinion on why milk from BST-treated cows should not be labeled. However, before coming to FDA, Dr. Margaret Miller was working for the Monsanto company as a researcher on BST. At the time she wrote the FDA opinion on labelling, she was still publishing papers with Monsanto scientists on BST. It appears to us that this is a direct conflict of interest to have in any way Dr. Miller working on BST. As you know, if milk is labeled as being from BST-treated cows, consumers will not buy it and Monsanto stands to lose a great deal of money. Several of Dr. Miller's former colleagues would lose their jobs."

The employees also complained that soon after Miller came to the FDA, she increased the allowable levels of antibiotics in milk from one part per 100 million to one part per million (ppm). Such a change was absolutely necessary in order to get rbGH approved. This is because farmers needed increased antibiotics to treat the increased udder infections. Injections of the medicines, however, end up in the milk at levels the FDA formerly considered unsafe—that is, until Miller changed the agency's criteria.

The anonymous letter continues, "She picked an arbitrary and scientifically unsupported number of 1 ppm as being the allowable amount of antimicrobial in milk permitted without any consumer safety testing." They pointed out that the 1 ppm limit is for each type of medicine. Milk could have many antibiotics at this level. "Effects of the different antibiotics could be additive and this is not taken into account." The letter claimed, "Dr. Miller's policy was used as the basis for approval of BST despite increased antibiotic usage. This also is a direct conflict of interest to have Dr. Miller working on this issue."

The letter further charged, "This is not the first time that CVM employees have charged Dr. Livingston with fraud and abuse leading to an endangerment of the public safety. However, it seems if anyone speaks out, they, not Dr. Livingston, end up in trouble. We as government employees cannot understand why it is allowed to continue."[6]

On April 15, 1994, three members of Congress responded to the allegations by asking the US General Accounting Office (GAO) to investigate. The congressmen asked the GAO, which conducts investigations within the government, to not only look into potential conflicts of interest by Margaret Miller, but also by Michael Taylor and Susan Sechen.

As Deputy Commissioner of FDA for Policy, Taylor (with support from Miller) determined that milk from rbGH-treated cows should not be labelled as such. He further wrote guidelines making it difficult for dairy producers even to label their milk as "rbGH free". Prior to joining the FDA, Taylor worked at a law firm, where, according to the congressmen's letter, "Monsanto was his personal client regarding food labelling and regulatory issues." That law firm uses Taylor's guidelines to sue dairies with rbGH-free labels. Sechen formerly conducted Monsanto-sponsored research before joining the FDA where she helped approve rbGH.

The congressmen wrote, "The entire FDA review of rbGH seemingly has been characterized by misinformation and questionable actions on the part of both FDA and the Monsanto Company officials." The letter also describes the previous attempt by the GAO to investigate the rbGH approval process, which they "had to abandon . . . because of the Monsanto Company's refusal to make available to them all pertinent clinical and related data."[6]

Evaluating the FDA's Evaluation

Growth hormones are naturally created in the pituitary gland of all animals. "It was known as early as the 1930s that injection of dairy cows with bovine pituitary extracts increased milk yield."[7] The practice, however, was not commercially viable until genetic engineering created a cost-effective production method. Engineers took the cow gene that creates their growth hormone, altered it and inserted it into *E. coli* bacteria, creating a living factory. The resulting hormone is similar, but not identical to the naturally occurring variety. When injected into a cow, it boosts the whole metabolism, including an increase in mammary cell activity. This leads to increased milk production.

Monsanto's growth hormone was controversial. It was the FDA's first look at a genetically modified food-related product. As such, there was a lot riding on its approval. Peter Hardin, editor of the Wisconsin-based dairy industry newspaper *The Milkweed*, says, "It was the lead off batter that *had* to get on base—because there was so much corporate influence behind it."[8]

Years after the drug was on the market, the Canadian scientists compiled a lengthy report that recounted all the various omissions, contradictions, weaknesses and gaps in the FDA's approval process. It came to be known as the Gaps Analysis Report, and charged that the FDA's "1990 evaluation was largely a theoretical review taking the manufacturer's conclusions at face value. No details of the studies nor a critical analysis of the quality of the data was provided."[9]

A look at the FDA's approval process of rbGH, which was perhaps the most controversial drug approval in FDA history, is telling. To defend the drug, the agency did something unprecedented. In August 1990, two FDA scientists published a paper in the highly respected journal *Science*, endorsing the product's safety. The authors referred to two studies in which rats were fed or injected with rbGH and then monitored for health changes. Although humans are not fed or injected with rbGH, cow's milk contains a small amount of naturally occurring bovine growth hormone (bGH). When cows are injected with rbGH, it is possible that some of this genetically modified version finds its way into the milk along with the natural variety. Feeding the rbGH to rats, therefore, was one way to see if the ingested GM hormone might lead to any problems.

The first of these studies lasted for only twenty-eight days. While the FDA claimed that rats fed rbGH showed no effects, the short duration chosen for the study astounded critics. Health Canada's Chopra says, "In twenty-eight days, what can you find?"[10] The second study was also short-term—ninety days.

According to the Canadian Gaps Analysis Report, since rbGH was a hormone, "its chemistry should have prompted more exhaustive and longer toxicological studies in laboratory animals." These are "usually required . . . to ascertain human safety." Because they weren't conducted, "such possibilities and potential as sterility, infertility, birth defects, cancer and immunological derangements were not addressed."[9] A study to determine whether a drug is carcinogenic will test two different species for about two years—the lifetime of mice or rats. Ignoring the short duration of the ninety-day study, the authors of the *Science* article pointed out that while the rats that were injected with rbGH showed significant changes, those fed rbGH showed no biological effects, proving the hormone to be safe if eaten.

After the rbGH was approved years later, Robert Cohen decided to analyze the conclusions in the *Science* paper. Although Cohen was a businessman, he had studied science as an undergraduate and worked at a laboratory—he was looking forward to running the numbers. He took the article and his calculator, and spent three days in his office pouring over the figures. He was unhappy at what he found. A lot of the data he needed was missing. Conclusions were often based on summarized data or on unpublished, company studies where the details were not in the public domain. In particular, while the authors indicated that the weights of many organs and tissues were measured at the end of the ninety-day feeding study, the *Science* article listed the data for only four of them.

Cohen called the FDA in 1994 asking to see the rest of the measurements. According to Cohen, the agency official, Richard Teske, told him it was a trade secret. Cohen was mystified why organ weights were considered a trade secret. Even the Inspector General of the Department of Health and Human Services, Richard P. Kusserow, had affirmed in a 1992 report that "complete disclosure of BST data will not occur unless and until FDA approves the drug for commercial use."[6] But in February 1994 Monsanto's rbGH was approved and now Cohen wanted complete disclosure.

Cohen tried to get it from Monsanto. No luck. He filed a Freedom of Information Act Request for the study, but was refused. He appealed within the FDA and, surprisingly, the request went to the *same* people who had initially refused him. Not surprisingly, Cohen lost again. The reason for the FDA's denial stated, "Release of the information would cause substantial competitive and financial harm to the company (Monsanto)."[6] Cohen was determined to get his hands on the study. He filed a lawsuit in Federal Court. The court ruled against him, again on the basis that the organ weights were a trade secret; revealing them, according to the ruling, could cause "competitive substantial harm"[6] to Monsanto.

(Contrast Cohen's experience with the statement made by US Speaker of the House J. Dennis Hastert on March 26, 2003. In defence of the safety of GM foods, he said, "Throughout the [approval] process, the public has ample opportunity for participation and comment, and data on which regulatory decisions are based are readily available.")

Relying on a friend's connection to Congressman Newt Gingrich, Cohen was given the opportunity to meet with officials at the FDA on

April 21, 1995 to discuss his concerns over the study. By this time, Cohen had heard that the FDA never actually received the raw data. He asked Richard Teske if the agency reviewed it. According to Cohen, Teske said they had reviewed it. But when Teske turned to Robert Condon who was in charge of statistical analysis, Condon admitted that they had, in fact, never reviewed it.

Similarly, three years later in October 1998, another FDA official, John Scheid, told the Associated Press that the agency had never examined the raw data from Monsanto's rat feeding study but rather based its conclusions on a summary provided by Monsanto.

This admission was serious. According to *Rachel's Environmental and Health Weekly*, "relying on a summary of a study, rather than on detailed data from the study, would violate FDA's published procedures."[11] In fact, the FDA scientists who wrote the 1990 article in *Science* stated that "the FDA requires the pharmaceutical companies to submit all studies they conducted on their products." Further, "The companies also submit the raw data from all safety studies that will form the basis of the approval of the product; the submission permits CVM scientists to confirm the accuracy of the results and conclusions."[7] In that same paper, however, the authors may have been basing their conclusions on research from the ninety-day rat feeding trial without having looked at the data.

When the Health Canada scientists did their investigation of rbGH, they determined that the FDA had "improperly reported" the data from the feeding study. The FDA had concluded that rbGH "was not and could not be absorbed into the blood stream". This conclusion was based on the assumption that "ingested rbGH would be expected to be degraded in the human gastrointestinal tract in the same manner as other proteins", and on the supporting evidence that there were no "clinical findings" among the rats fed rbGH.[7] Both were wrong. There are other proteins that do survive digestion and, according to the rat study data, there were clinical findings. In fact 20 to 30 percent of the rats fed rbGH developed an antibody response. Such a response, according to *Rachel's*, "is evidence that the immune system has detected, and responded to, a substance entering the body."[11] According to Chopra, it was apparent that the FDA never reviewed the antibody data. Furthermore, the Canadian team discovered that some male rats which were fed the hormone developed cysts on their thyroid and changes in their prostate gland, which should have prompted further investigation.

The official report from the FDA admits that they had accidentally overlooked the data in the antibody study, but, contrary to Scheid's comments to a reporter and what Cohen recalled from his meeting, they claim to have reviewed other data from the study.

In addition to the effects that ingested rbGH might have on humans, Chopra was particularly concerned about how sick cows might adversely affect the composition of milk and meat. The Canadian report said that the "numerous adverse effects [on the cows] . . . may have had an impact on human health",[9] and should have been taken into consideration by both the FDA and the Human Safety Division of Canada's Bureau of Veterinary Drugs (BVD).

For example, the Canadian Gaps Analysis Report pointed out that cows injected with rbGH did suffer from "birth defects, reproductive disorders, higher incidence of mastitis [udder infection]"[9] and other problems. Other sources report high incidence of foot and leg injuries, metabolic disorder, uterine infections, indigestion, bloat, diarrhoea, lesions and shortened lives. Six months before the *Science* article appeared, *The Milkweed* had published data that had been stolen from the FDA. It revealed that cows treated with the hormone for about eight months had major increases in the size of their hearts, livers, kidneys, ovaries and adrenal glands. In their report to the FDA, however, Monsanto dismissed the changes as "harmless physiological shifts".[12]

Although organs increase in size, the flesh decreases. According to Hardin, on average, the metabolic stresses placed on the treated cows will cause them to lose more weight than untreated ones during their milking cycle. Some treated cows are so lean and wasted by the end of their lives, they offered little value to slaughterhouses that normally convert the cows' carcasses into meat. The slaughterhouses also complained that the cows' tissue at the injection site was killed, sometimes leaving a swollen mound. It would have to be cut out of the meat before it went on the market.[8]

The stolen data also revealed that rbGH cows had more difficulty getting pregnant. While 95 percent of untreated cows became pregnant during the eight-month trial, only 52 percent of the treated cows did. According to *The Milkweed*, "Monsanto notes that pregnancy rates were not lower than rates normally observed in the dairy industry. That's statistical bunk. Reproduction data from that test shows Monsanto counted as pregnant many test cows which were pregnant before the treatments began!"[12]

Note: Monsanto's own product label warns of many health problems for cows that may occur when treated with rbGH. It also mentions that the milk might contain an increase in somatic cell counts due to increased infections. Somatic cell counts is another term for pus, or dead white blood cells. It is interesting to note that in order to facilitate the hormone's approval, Monsanto released research summaries describing pus content. But they only released summaries of eight small studies, some showing changes, some not statistically significant. Three independent scientists were able to obtain the data and derived the figure of an average increase of 19 percent. They also discovered that the percentage of cows in the high-pus category was about double what Monsanto had indicated, an important correction since farmers are charged penalty payments if specific pus levels are exceeded. When the three scientists tried to publish their research, they were thwarted by Monsanto, who successfully kept it out of the scientific journals for three years—until after rbGH was approved. (Even the 19 percent figure was low, since Monsanto had apparently dropped data from the last weeks of the study, when pus content was considerably higher.)

Hormones in Your Milk

The FDA did not appear to share the Canadian scientists' apprehensions about the effects that all of the cows' health problems may have on milk and meat. The FDA did, however, show concern about the dramatic changes in blood hormone levels immediately after injection. According to the stolen data from FDA files, hormone levels skyrocketed during Monsanto's test of six different rbGH formulations. In one group, for example, the bGH in their blood jumped to about 1,000 times that of the controls after treatment. To protect public health, starting in the early 1980s the agency mandated that milk from treated cows could not be sold until at least five days after injections. Slaughtering treated cows for meat likewise required a fifteen-day waiting period.

These requirements presented a serious problem for Monsanto. With cows receiving two injections per month, discarding milk for ten days each month would make the drug uneconomical. In fact any mandatory period where milk had to be thrown away would be an issue. As Hansen points out, "you can talk all you want about how safe it is, but if there are pictures of milk being dumped, the obvious question is: 'Why can't that milk be used, why is that being dumped?'"[13] Therefore, in February 1985,

Monsanto requested that the FDA allow their milk to be sold without a "withdrawal" period. They presented their twenty-eight-day rat feeding study as rationale. The FDA went for it. The FDA also removed the waiting period for slaughtering cows. In 1988, the FDA's Susan Sechan sent a letter to manufacturers asking them to measure the hormone levels in the blood. They said no. According to *The Milkweed*, "Replies indicate that the industry believed testing blood of treated cows for added hormones would raise public health/safety issues. Sechen did not press for those blood hormone tests to be conducted."[12]

In fact, the FDA didn't even require tests to see how much rbGH residue, if any, enters the milk. If rbGH is present, however, it does not mean that humans will have the same health problems that the rats and cows experience. The residue would likely be a small fraction of the amount given to these animals. But we don't know how small, since the agency granted Monsanto an exception to their normal requirements by allowing the drug to enter the market without also requiring that an assay test be available to measure residues.

Instead, scientists rely on tests that measure overall bGH levels. This includes naturally occurring bovine growth hormone (bGH), as well as any of the injected drug (rbGH) that might happen to get mixed in. While cows injected with extra high doses of rbGH showed increased amounts of bGH in their milk, the *Science* paper reported that with normal doses, "they found that levels of bGH in milk from rbGH-treated cows (4.2 ng/ml) were not significantly different from those found in non-treated cows (3.3 ng/ml)."[7] The details of the study, however, reveal that it may have been designed specifically to force this conclusion.

The research was conducted on only three cows. Using so few cows makes it easier to label moderate changes in bGH levels as not significant. Thus, the 26 percent increase noted by the *Science* article was deemed "not significantly different". The researchers used a version of rbGH that was manufactured by American Cyanamid—different in composition from the Monsanto version that was eventually approved. More importantly, the study treated the three cows with 10.6 mg of rbGH per day. The *Science* paper describes this as "approximately the proposed dose".[7] Although American Cyanamid might have specified a dose of approximately 10.6 mg per day as an option, Monsanto's rbGH is not administered on a daily basis. That would be too labour intensive and impractical. Rather, cows are injected with a larger dose—500 mg—every two weeks. That amount is forty-seven times the 10.6 mg used each day in this study. We know from the stolen data described above that when cows are injected with the higher dose every two weeks, the bGH in their blood skyrockets. The bGH in their milk might be similarly elevated. By selecting a small, daily dose format, however, the researchers in this experiment would avoid the large post-injection jump, thus skewing the results to show no significant difference in the milk. The lead researcher on the bGH study was Paul Groenewegen, an undergraduate student; his three co-authors all had close ties with Monsanto, having published research with Monsanto scientists or conducted research funded by the company.

The *Science* authors assure us, however, that "the need to pursue more definitive studies has already been stated as unnecessary because bGH is biologically inactive in humans and orally inactive."[7] As for the claim that it is orally inactive, the Canadian scientist showed that to be false based on the rats' antibody reactions. The FDA's claim that bGH is inactive is derived from their assumption that it is a species-specific hormone; that it works in cows but not in humans. As supporting evidence the scientists say that the molecule's amino acid sequence differs from the human growth hormone (hGH) by about 35 percent. The *Science* article cites studies in the 1950s showing that while human dwarfs who were injected with hGH grew taller, those injected with bGH did not. Hence, it doesn't have an effect on humans.

A study published in 1965, however, demonstrated that bGH *does* have an effect on humans. For people with pituitary deficiency, bGH treatments produced some changes "similar to those effects noted after administration of hGH." The FDA should have known this, as this quote appeared in a new drug application submitted to the FDA in 1987 by Elanco—for their brand of rbGH. Another study also showed other minor changes in human physiology resulting from bGH.

Irrespective of whether bGH levels increase, influence humans, or are mixed with the genetically modified variety in milk, the FDA assures us that there is still no problem. The *Science* article says, "It has also been determined that at least 90 percent of bGH activity is destroyed upon pasteurization of milk. Therefore, bGH residues do not present a human food safety concern."[7] Cohen decided to investigate this claim. It took him some time to track down the source for the pasteurization information, as it had been improperly cited in the *Science* article. It turned out to be part

of the same Groenewegen study described above. The paper had been published just two months before the article in *Science* appeared. (Some
observers believe that the submission to *Science* was held up, waiting for
the Groenewegen study to appear.)

The paper described how they heated milk at 162°F for thirty minutes.
Cohen said, "when I read that—I said, wait a second, milk is pasteurized
for fifteen seconds at that temperature—not thirty minutes. They intentionally tried to destroy the hormone."[14] "That must have been their mission. Why else would they heat the milk for thirty minutes at a high
temperature reserved for a fifteen second treatment?"[6]

Hardin points out, "The difference between fifteen seconds and thirty
minutes is 120 times. From a dairy standpoint, milk pasteurized for that
long would have no nutritional value. What they did is the worst of bad
science." Hardin illustrates his point with a graphic analogy: "If you cook
a turkey 120 times longer than the recommended time, and then you tried
to extrapolate scientific conclusions based on the charred remains, you'd
be thrown out."[8]

But even after thirty minutes, only 19 percent of the bGH in milk from
hormone-treated cows was destroyed. According to Cohen, "They then
'spiked' the milk. This is their word, 'spike'. They added artificial BST . . .
146 times the level of naturally occurring BST in powdered form to the
milk and heated it. The powdered BST in milk was destroyed! They saved
the day for Monsanto. The experiment worked. These men of science could
claim that heat treatment destroys BST. In their concluding discussion,
these scientists determined: Heat treatment effectively reduced the
immunoreactive quantities of BST in milk."[6]

Insulin-Like Growth Factor

Most critics of rbGH are far less concerned about rbGH residues in the
milk than they are about another change that takes place in treated cows.
The injections result in the increase of yet another hormone: insulin-like
growth factor-1 (IGF-1). Humans also have IGF-1. It causes cells to divide
and is one of the most powerful growth hormones in the body. It also
resembles insulin; hence the name insulin-like growth factor. Human IGF-
1 is chemically identical to the IGF-1 found in cows. Since rbGH-treated
cows have more IGF-1 in their system, it is vital to know if their milk also
has increased levels of IGF-1. If it does, will it be absorbed into our system and how will that affect us?

Monsanto researchers Robert Collier and others tried to assure the public that IGF-1 wasn't an issue. In a letter published in the *Lancet* in 1994, they wrote, "IGF-1 concentration in milk . . . is unchanged" and "there is no evidence that hormonal content of milk . . . is in any way different."[15] A month later, however, a letter in the same publication from a British researcher "reminded Monsanto that in its 1993 application to the British government for permission to sell rbGH in England, Monsanto itself reported that "the IGF-1 level went up substantially.""[16]

Even the FDA admits, "rbGH treatment produces an increase in the concentration of insulin-like growth factor-1 (IGF-1) in cow's milk."[7] The amount of increase is disputed, depending on the study. While some supporters of rbGH acknowledge "that it at least doubles the amount of IGF-1 hormone in the milk",[17] the first study on the subject reported an increase of 360 percent.[18] Margaret Miller's research demonstrated an increase of 47 to 71 percent. Whatever the amount, according to *Rachel's Environmental and Health News*, "IGF-1 in milk is not destroyed by pasteurization."[19] It's intact in the milk we drink.

Not only is there more IGF-1 in the milk, it is not destroyed in the stomach. The Canadian Gaps Analysis Report said IGF-1 "can survive the GI tract environment" and is "absorbed intact". The report says, "The full significance of this finding also was not investigated [in the FDA's evaluation]."[9]

The amount of IGF-1 that gets absorbed may be much higher due to the fact that it is mixed in with milk. Using radioactively labelled IGF-1, Japanese researchers reported in 1997 that while only 9 percent of IGF-1 fed to mice ended up in the bloodstream, the figure jumped to 67 percent when the hormone was fed along with casein, the major protein in milk. This buffering effect of milk helps explain why breast milk is nature's method for delivering absorbable hormones to infants.[20] (This raises the question about whether rbGH would be similarly buffered by milk. While the ninety-day rat feeding study discussed earlier showed that rats that were fed rbGH had much less severe health effects than those rats that that were injected with the hormone, perhaps the difference would not have been as great if the rats were fed the hormone with milk.)

A study on humans published in the *Journal of the American Dietetic Association* measured the free levels of IGF-1 in two groups, one that drank milk and one that did not. The study reported, "Serum IGF-1 levels increased significantly in milk drinkers, an increase of about 10 percent above baseline, but was unchanged in the control group."[21]

Although some amount of IGF-1 is necessary, elevated levels of the hormone have been linked with cancer. By 1991, it was already documented that IGF-1 was "critically involved in the aberrant growth of human breast cancer cells".[22] That year, the Council on Scientific Affairs of the American Medical Association called for more studies to determine if ingesting "higher than normal concentrations of [IGF-1] is safe for children, adolescents and adults".[19] In 1993, the European journal *Cancer* concluded that IGF-1 "plays a major role in human breast cancer cell growth".[23]

In January 1998, a paper in *Science* established the cancer link even further. A Harvard study of 15,000 white males revealed that those with elevated IGF-1 levels in their blood were four times more likely to get prostate cancer than the average man. The report says, "A strong positive association was observed between IGF-1 levels and prostate cancer risk" and "administration of GH [growth hormone in humans] or IGF-1 over long periods . . . may increase risk of prostate cancer."[24]

Four months later, a study in the *Lancet* found pre-menopausal women in the US with high levels of IGF-1 were *seven* times as likely to develop breast cancer. The authors wrote, "with the exception of a strong family history of breast cancer . . . the relation between IGF-1 and risk of breast cancer may be greater than that of other established breast-cancer risk factors."[25]

In January 1999, the *Journal of the National Cancer Institute* reported, "IGF-1 strongly stimulate[s] the proliferation of a variety of cancer cells, including those from lung cancer. High plasma levels of IGF-1 were associated with an increased risk of lung cancer."[26] *The International Journal of Cancer* described the "significant association between circulating IGF-1 concentrations and an increased risk of lung, colon, prostate and pre-menopausal breast cancer", and concluded, "Lowering plasma IGF-1 may thus represent an attractive strategy to be pursued."[27] Milk from rbGH-treated cows may be doing just the opposite.

Cancer Research reports, "Diet contributes to over one-third of cancer deaths in the Western world, yet the factors in the diet that influence cancer are not elucidated."[28] Progress was made on this when research published in September 2002 explored the link between IGF-1 levels and the foods that people eat. Scientists looked at data from more than 1,000 nurses who carefully recorded their diet. After analyzing dozens of food categories, the researchers concluded that the food most associated with high IGF-1 levels is milk. Michelle Holmes, who led the study, said, "This association raises the possibility that diet could increase cancer risk by increasing levels of IGF-1 in the blood stream." Their data also revealed that women who have had multiple pregnancies had an average of 15 percent less IGF-1. "The finding," reported Reuters, "could explain why women who have had children have a lower risk of cancer—something doctors have noticed but [have] been unable to explain."[29] The research, which was conducted by a team from Brigham and Women's Hospital and Harvard Medical School, did not look at milk from cows treated with rbGH—the data was taken before it came on the market. But because treated milk has higher levels of IGF-1, many expect its effect on human IGF-1 levels to be greater.

Global Reactions to Canada's Revelations

Concerns about IGF-1, like many of the other issues raised in the Gaps Analysis Report, was a frustrating setback to many Senior Health Canada officials who were determined to get the rbGH on the market. Their original intention was to approve the drug without review, simply because the US had. According to Chopra, in 1997 Director General George Paterson of the Food Directorate of Health Canada was in Geneva, two days before announcing its approval at an international conference, when the initial objections by the Health Canada scientists made headlines. The National Farmer's Union of Canada, which had been lobbying against the approval of rbGH in Canada, contacted the organizers of the meeting that Paterson was attending. This created pandemonium in the group that was being asked to approve it. Unable to proceed, the meeting was postponed for several months in order to allow more evidence on human safety to be investigated. An angry director general returned to Ottawa. When Chopra and his colleagues were called into his office, Chopra says, "He was pounding on the table saying that this was no way to hear about this issue."[10] Chopra stood his ground, saying that if the director general wanted to approve rbGH, of course he could; but the director general did not speak for him. Chopra said he would continue to blow the whistle on the drug of questionable safety if it was approved without proper review. Eventually, Chopra was asked to lead that review, which was undertaken by the entire staff of the human safety division.

The team was finally able to examine data from the files, which had been kept in the control of one senior staff member. When they finished their report condemning the FDA's approval of rbGH, their department head insisted that it be changed. He didn't want anything that incriminating of the drug to possibly get into the hands of the public. In fact, when Chopra and others were to testify before the Canadian Senate committee in charge of investigating rbGH, Chopra's boss told him that they could only submit the edited version of the report and not speak about the omitted portions. But the original version had already been submitted to a labour board that was investigating the scientists' reports of harassment—it was leaked to the US press in late 1998. The Canadian Senate committee demanded to see the report, which became news around the world.

It had an impact on an already controversial issue. The EU, which had a wait-and-see moratorium in place, ended up instituting a permanent ban. New Zealand, Australia, Japan and other industrialized countries likewise do not allow rbGH.

In the United States, the Gaps Analysis Report revelations prompted groups around the nation to challenge the FDA's approval. Two citizen activist groups, the Vermont Public Interest Research Group and Rural Vermont, charged in a joint news release that "Either the FDA or Monsanto covered up the results of the major human safety test."[3] The Washington, D.C.-based Center for Food Safety filed a request with the FDA to withdraw or suspend use of rbGH. According to *The Capital Times* of Madison, Wisconsin, Andrew Kimbrell, director of the center, "said the FDA should have known the hormone survives digestion and called it unconscionable that this information has been hidden from the public, which for the past five years has been consuming rbGH-treated milk."[30] Monsanto had, by that year, sold more than 100 million doses. Vermont's Senators Patrick Leahy and James Jeffords asked Secretary of Health and Human Services, Donna Shalala (who, incidentally had been pictured with a milk mustache as part of the dairy industry's ad campaign) to investigate the FDA's approval and see if it had overlooked key evidence. The FDA responded with a report. Although they admitted that their original safety assessment had failed to review the antibody portion of the rat study, the agency re-affirmed the safety of rbGH. It remains on the market.

The irony of the approval of rbGH in the US is that the White House had defended the hormone by saying that milk production would increase and prices would go down. But in 1986–1987, the government paid farmers to stop dairy farming for five years and even to kill more than 1.5 million dairy cows in an attempt to stop the overproduction of milk and to boost prices.[31] The Office of Technology Assessment (OTA) did not share the White House's enthusiasm for rbGH. A May 1991 OTA report said, "If approved by FDA, [it] will accelerate trends that already put additional economic stress on dairy farm operators in many areas of the country."

The White House offered another, perhaps more honest, reason why it supported rbGH approval: "US leadership in biotechnology, as well as private-sector investment for research and development in the biotechnology industry, would be enhanced by proceeding with BST, and would be impeded if there were new government obstacles to such biotech products."[6]

Corporate Influence on Government

Government's allegiance to industry is not uncommon. The FDA's history, in particular, is replete with evidence demonstrating their loyalty to industry at the expense of human health. In 1960, Senate investigators discovered that FDA officials had been receiving financial incentives from the companies they were supposed to regulate. The director of the FDA antibiotic division, for example, received $287,000 in "honoraria". In 1963, John Nestor, a pediatrician with the FDA, told Senate investigators that the FDA "worked too closely with the giant drug companies to be effective."[32] In 1969, a congressional study revealed that thirty-seven of forty-nine top FDA officials who left the agency took jobs with food and drug companies.[33]

In 1976, legislative investigators indicted the agency and thirty-four of its current and former key employees for "failing to fully protect the public". The scandal involved the FDA, pharmaceutical manufacturers, doctors and research scientists who were said to unnecessarily expose humans to risks in the testing of new drugs. Also that year, the GAO reported that 150 FDA officers had violated federal "conflict of interest" rules by owning stock in drug companies the agency was monitoring. A House committee charged that the FDA's "advisory committees" were subject to "improper influence" from drug manufacturers.[32]

Several FDA officials and drug company executives were convicted on corruption, racketeering and similar charges for a bribery scheme that went on from 1989–1992. Generic drug companies paid off FDA officials to approve their drugs and block approval of competitors' drugs. The generic drug companies also withheld data and even substituted other companies' brand name drugs for evaluation, instead of risking an evaluation of their own product.[33] The Department of Health and Human Services said that the FDA's internal controls were inadequate to insure integrity in the drug-approval process, leaving the agency open to "manipulation and preferential treatment".[32]

One apparent example of preferential treatment was revealed in a 1993 testimony before a House subcommittee, describing the approval process of GD Searle's aspartame (NutraSweet®)*, an artificial sweetener produced by genetic engineering. Between 1977 and 1983, top White House officials, two former FDA commissioners and three ranking agency officials, all used their influence to get aspartame on the market. In 1990, more than 5,500 consumers filed complaints to the FDA describing adverse reactions to this sweetener. That accounted for 80 percent of all complaints about a food or additive for the whole year. When a 1996 study statistically linked it with human brain cancer, the FDA and Monsanto, which had purchased Searle in 1985, "were quick to assail the information and reaffirm the product's safety".[32] The FDA's lead spokesperson to defend the product became vice president for clinical research at Searle two years later.

Former FDA Commissioner Dr. Herbert Ley said in 1969, "The thing that bugs me is that people think the FDA is protecting them. It isn't. What the FDA is doing and what the public thinks it's doing are as different as night and day."[34]

The Canadian regulatory agencies appear to have some of these colourful characteristics of their own. When the Canadian Gaps Analysis Report was made public, Monsanto spokesman Gary Barton "said the safety of rbGH will be 'reaffirmed' in the coming weeks when two Canadian peer review panels, empowered by Health Canada, are expected to release their findings."[30] Barton's confidence in the outcome of the two Canadian review panels was telling.

Health Canada sometimes uses outside panels when its own scientists cannot reach a consensus. But, according to a 1998 Council of Canadians report, "the scientists *did* reach a consensus"—they did not want the hormone approved. Internal documents obtained through Canada's Access to Information laws indicate that long before the scientists were even preparing their report, "Health Canada officials were preparing to defend the rationale for [using] external panels" in order to force an approval. They

---

* NutraSweet® is a registered trade mark of NutraSweet Property Holding Inc.

selected one panel for reviewing animal health and another for human health.

Health Canada's policy states that all members of external panels must not only be free from actual conflicts of interests, but also "must not have material interest in the result" and "not create a reasonable apprehension or suspicion of bias". The rules notwithstanding, the Council of Canadians reported that both panels included members that have close ties to Monsanto and the industry. One member worked for Monsanto, "one receives funding from a company with a profit-sharing agreement" with Monsanto for rbGH-related products, and the spouse of the chairman of one committee worked for GD Searle, a subsidiary of Monsanto for fifteen years. "Panel members have spoken in favour of [rbGH], or on closely related subjects." And one of the two panels is actually sponsored by "an organization whose support for rbGH is a matter of public record".[35]

In spite of the close ties to industry and the confidence that Monsanto had in them, the issue apparently became too controversial to merit an approval. In January 1999, Health Canada announced that it would continue its decade-long ban on rbGH. It did not, however, acknowledge that there was a human health issue. The panel looking at human health issues, which had never been given the full Gaps Analysis Report and which did not meet with its authors, was not the group that blocked the drug's approval.

According to Canadian Broadcasting, the panel made up of veterinarians that was looking at animal safety "concluded that the hormone would be too dangerous for cows". They admitted that the hormone "causes an udder infection called mastitis, infertility and an increased risk of lameness. The problems could be so severe," CBC said, "farmers risked having to destroy up to a quarter of their herds."[36] This was particularly problematic in Canada. Unlike the US, where rbGH was available as an over-the-counter drug, in Canada it was to be offered by prescription only. Furthermore, veterinarians are the ones to prescribe it, and could be sued whenever adverse reactions took place. Thus, the committee reviewing animal safety was also looking after the economic safety of the nation's veterinarians.

The rbGH controversy, however, did not end there. On December 7, 1999, Health Canada scientists told a new Canadian Senate Committee how they were threatened, harassed and denied promotions in retaliation for their testimony the previous year.[37] When the committee learned that scientist Shiv

Chopra had been suspended for testifying, they asked to see his boss, Andre Lachance. But Lachance disappeared a few days later. His attorney claimed he was sick and unable to appear before the committee. The department then replaced Lachance altogether and announced that he wouldn't be returning at all. The Senate Committee described the situation in the department as "deplorable".

Sour Milk

Leading up to the approval of rbGH, the land grant colleges in the United States, which receive significant contributions from the biotech industry, heavily promoted it to dairy farmers. Using their network of extension agents, the prevalent message was that if a farm didn't adopt it, they would likely become insolvent. One paper from Cornell University, entitled "The Impact of BST on Dairy Farm Income and Survival", estimated that farms that didn't use rbGH would lose between $6,000 and $20,000. The paper said, "those who adopt early and obtain good production responses will find their rewards great."[38] According to another Cornell University study done from 1994 to 1997, however, farms that used rbGH did not actually achieve higher profits than farms that did not.[39]

In spite of the lack of profitability, in 2002, more than 2 million of the 9.2 million US dairy cows were injected with rbGH. Larger dairies farms use it more often; 54 percent of farms with at least 500 cows compared to about 9 percent of farms with less than 100 cows. But since dairies typically mix milk from many farms, milk from hormone-treated cows is in almost all US dairy products. The drug is more popular among farms in the West, and its use is rising nationwide.[40] Monsanto reported an 8 percent increase in sales in 2002.

Settepani says the FDA's failures "have become a scientific breakdown that threatens the safety of America's food supply generally, and dairy products in particular. . . . As a result of [their] failure to act, the nation's milk supply—as well as products such as infant formula, ice cream, cheese and yogurt that are made from milk—is highly contaminated with unknown levels of animal-drug residues that have not been shown to be safe."[4] Each year, Americans consume nineteen gallons of milk, thirty pounds of cheese and four pounds of butter, as well as dairy in other forms. Taken together, these require sixty-five gallons of milk per person.

Organic dairies and many others take precautions to avoid the hormone. Oakhurst dairy of Portland, Maine, for example, requires its suppliers to sign a notarized affidavit every six months stating that they won't use it on their herds. The small dairy pays them an extra $.20 per 100 pounds of milk—totalling half a million dollars in 2002. Their label states, "Our Farmers' Pledge: No Artificial Growth Hormones."

In early July 2003, Monsanto filed a lawsuit against Oakhurst Dairy, claiming that their labels "deceive consumers". Monsanto's spokesperson said, "They're marketing a perception that one milk product is safer or of higher quality than other milk. Numerous scientific and regulatory reviews throughout the world demonstrate that that's unfounded."[41]

Earlier in 2003, Maine's attorney general refused Monsanto's request for the state to stop using its Quality Trademark Seal programme, which is used by dairies to indicate when milk is free of artificial growth hormones. In their argument against the programme, Monsanto said that the label should appear in the proper context with the following language, suggested by the FDA: "No significant difference has been shown between milk derived from rbST-treated and non-rbST-treated cows."[42] These words were written by Michael Taylor, the attorney who represented Monsanto before becoming an FDA official, and who was later hired by Monsanto as a vice president.

Wisdom of the Cows and Pigs

Bill Lashmett watched as two or three cows were let into a feeding area at a time. The first trough they came to contained fifty pounds of shelled Bt maize. The cows sniffed it, withdrew and walked over to the next trough, which contained fifty pounds of natural shelled maize. The cows finished it off. When they were done and released from the pen, the next group came in and did the same thing. Lashmett said the same experiment was conducted on about six or seven farms in Northwest Iowa, in 1998 and again in 1999. Identical trials with pigs yielded the same results, also for two years in a row.

Lashmett, who has a background in biochemistry and agriculture, says that animals have a natural sense to eat what is good for them and avoid what isn't. He witnessed this first-hand in another experiment conducted by a feed store in Walnut Grove, Iowa. They put twenty-three separate vitamins and minerals, each in their own bin, out where cows could eat them. The cows would alternate their choice of bins in such a way, according to Lashmett, that they received a balanced, healthy diet. Moreover, their preferences changed with the seasons and climate, demonstrating a natural inclination to follow the dictates of their bodies' needs.[1]

Chapter 4

Deadly Epidemic

Betty Hoffing, at age sixty-one, was a bit of a local character in her hometown of Skokie, Illinois. A travel agent for twenty-five years, Betty's infectious humour and unbounded energy made her a favourite not only in the travel industry but also in the volunteer organizations where she spent several after-work hours each week. She was in excellent health and never had a serious health problem in her life. Not until August of 1989.

One day at work, Betty was suddenly overcome with intense pain in her chest and down her left arm. Her doctor, thinking it was a heart attack, had her go immediately to the intensive care unit of a nearby hospital. But two days later, after batteries of tests came up with nothing, the doctors sent Betty home. There was no heart attack, no explanation.

The next month, she had developed a mysterious rash all over her body. Soon after came a horrible cough. By the end of September, Betty was hit with the worst symptoms so far—severe muscle weakness and extreme pain. "It was hard to walk, hard to do anything,"[1] she said. Her muscles were going haywire. Without warning, her hand or jaw would close shut; any muscle would just lock down. If she were driving at the time, she'd have to quickly pull over and wait out the painful spasm. Her physicians were baffled.

Betty was forced to take a leave of absence from work. In mid-November, she decided to spend the day in bed. She didn't leave that bed for nearly six months. Her pain was so severe, that even rolling over was unbearable—it took her two full minutes.

* * * * *

One day in March 1989, Harry Schulte, a Catholic deacon living in Cincinnati, was sitting in front of his television when all of a sudden he heard what sounded like a shotgun go off in his head. "I thought I was going crazy," he recounted. He wasn't. He was experiencing the first symptom of a disease that would turn his life upside down.

Within weeks, the nightmare began. "I would sit up on the side of the bed and try to sleep sitting up because of the intensity of the pain. My legs became—you wouldn't believe it unless you saw it—they became as big as a telephone pole. They split and water oozed from them. No amount of medicine they gave me . . . calmed the pain."[2] Schulte would eventually lose his job and his family.

* * * * *

In the summer of 1989, Janet O'Brien of California was struck. At its worst, the pain was so severe she "could barely stand to be touched". She says, "I lost about 60 percent of my hair, had no energy, and was usually asleep. At various times, I have had mouth ulcers, nausea, shortness of breath, severe muscle spasms, itching and painful rashes all over, edema (swelling of extremities), concentration and memory difficulties, hand-writing problems, balance problems, irritable bowel syndrome, weight gain, visual perception problems, just to name a few symptoms!"[3]

* * * * *

All over the US that year, patients like Janet, Harry and Betty were struck with mysterious debilitating symptoms. For many, the pain was greater than their doctors had ever seen.[4] Some also experienced harden-ing of the skin, cognitive problems, headaches, extreme light sensitivity, fatigue and heart problems. The worst cases were crippled by "ascending paralysis, in which a person loses nerve control of the feet followed by the legs, then bowels and lungs, finally requiring a respirator in order to breathe."[5]

Doctors were mystified. There was nothing in the medical literature to explain this illness and no treatment was stopping it or slowing it down. To make matters worse, no one yet knew it was an epidemic. The symp-toms varied widely and the outbreak was dispersed—doctors would gen-erally see only one case.

This was true of Phil Hertzman, a medical doctor from Los Alamos, New Mexico. In October 1989, Kathy Lorio, a forty-four-year-old woman who had been healthy and strong came to see Hertzman after she was sud-denly hit with debilitating pain and other serious symptoms. After running tests, Hertzman noticed something in her blood that stopped him cold. The normal count for the type of white blood cells called eosinophils is around 10 per cubic centimetre. For patients with an allergy or asthma, these can rise to 200 or 300, even 500. Hertzman's patient was off the charts. Her count was about 10,000.

Hertzman referred Lorio to Santa Fe rheumatologist James Mayer. Although Mayer could not find a cause for her pain, he happened to have recently seen a second patient, Bonnie Bishop, also with severe pain, muscle weakness and a high blood count. In addition, "Bishop's arms and legs had filled with fluid and her breathing was laboured. When she tried to sit up, she slumped like a rag doll because her back muscles were so weak."[6] Mayer could not find a cause for Bishop's symptoms either, but she had given him a list of all the supplements she had been taking. Mayer asked if Lorio had been taking any of these. When he asked about L-tryptophan, there was a match. Lorio was taking it to help her sleep.

The doctors phoned Gerald Gleich at the Mayo Clinic, an internationally renowned expert on eosinophils. They told him about the L-tryptophan. But two cases weren't enough to draw a conclusion, Gleich said. Better wait. They didn't wait long. That same day a third case, also linked to L-tryptophan, was reported in New Mexico. Gleich called the Center for Disease Control in Atlanta and told them about the link to L-tryptophan.

L-tryptophan is an amino acid, a building block of plant and animal proteins. It is one of the "essential amino acids", those that need to be supplied from the diet since it is not manufactured in the body in adequate amounts. L-tryptophan aids in the production of serotonin, which promotes sleep. The presence of L-tryptophan in milk and turkey explains why these foods have been associated with helping people sleep or relax. L-tryptophan was available as an over-the-counter supplement recommended by doctors and others for "insomnia, premenstrual tension, stress and depression".[7] As L-tryptophan had been used safely for years, the doctors were not yet sure if it was the cause of their patients' disorders. Furthermore, all three cases were from New Mexico. Perhaps there was some local toxin that was the cause.

Within two weeks, Gleich learned from colleagues that three more patients apparently with the same disease had checked into the Mayo clinic. They were from different parts of the US. One was already on a respirator in very serious condition. All three had taken L-tryptophan. Gleich phoned the CDC again. He told them that the disease was not limited to New Mexico—it's out and it's deadly.

In the meantime, Tamar Stieber of the *Albuquerque Journal* became aware of the mysterious disease and its potential link to L-tryptophan. On

November 7, in the first of a series of articles that would eventually win her a Pulitzer Prize, she described the disease and the possible cause. Immediately, the calls starting pouring in: others who had taken L-tryptophan were also reporting similar symptoms.

Four days after the article appeared, the FDA sent out "a strong warning to the public"—stop using L-tryptophan! Within days another 154 cases were reported from around the nation. The FDA responded by issuing a recall—all over-the-counter supplements containing 100 mg or more of L-tryptophan were to be removed from the market. The level of 100 mg was chosen, according to FDA testimony, "because, at the time, the lowest daily intake associated with illness was 150 mg."[7]

The CDC named the disease eosinophilia-myalgia syndrome or EMS. It was named for the high number of eosinophils (eosinophilia) and for the severe muscle pain (myalgia). By early December, the reported EMS cases jumped to 707, with one death linked to the outbreak and several others under investigation. By late March, the number of cases had reached 1,411, including 19 deaths. Although CDC stopped monitoring the disease shortly after resolution of the epidemic, the agency's final estimate put the number of cases between 5,000 and 10,000[8] and the number of deaths near forty. A recent but incomplete survey among a 1,000-member network of EMS victims in the US suggests that as many as 80 to 125 people with the disease have died. But it is difficult to assess how many, and to what degree, these fatalities were impacted by EMS.

In March 1990, responding to the report that one person had contracted the disease after taking a dose below 100 mg, the FDA expanded the recall to all over-the-counter L-tryptophan. The FDA waited almost another year to recall some forms of L-tryptophan prescribed by doctors, such as those used in intravenous administration and infant formula.

Tracking the Source of the Outbreak

Only six manufacturers, all Japanese, supplied L-tryptophan to the US. After months of investigation, CDC researchers concluded, "only L-tryptophan manufactured by Showa Denko KK was clearly associated with illness."[9] Showa Denko K.K. was Japan's fourth largest chemical manufacturer and the largest supplier of L-tryptophan to the US market.

When researchers analyzed Showa Denko's L-tryptophan, it "was found to have much higher levels of impurities than other manufacturers' product."[9] There were sixty to sixty-nine trace contaminants found in their L-tryptophan, six of which were associated with EMS cases. Although the contaminants were tiny, measuring as little as 0.01 percent (1 part in 10,000), scientists believed that one or more of these six were responsible for the outbreak of disease.

According to Showa Denko's attorney, Don Morgan, "There is no evidence to suspect that any external materials got into the production process and 'contaminated' the product. The manufacturing process was carefully controlled."[9] If the impurities didn't come from outside, where did they come from, and why were they only found in Showa Denko's product?

To produce L-tryptophan, most of the Japanese manufacturers combined certain strains of bacteria and enzymes in a fermentation process. The resultant fermentation "broth" passed through a filter to purify the product. Showa Denko, however, had pioneered a new method of production: They genetically engineered their bacteria to dramatically increase yields. This strategic move, however, carried greater risks.

According to Charles Yanofsky, professor of biology at Stanford University, when "higher than normal concentrations of certain enzymes and products" are produced through genetic engineering, it could end up creating "higher levels of toxic substances". He says, "Anytime you overproduce a small molecule in bacteria, there are unknowns of this type."[9]

Yanofsky, a leading expert in L-tryptophan biosynthesis, says Showa Denko's bacteria probably produced ten to fifteen enzymes and other byproducts in excess of normal levels. If other enzymes, in turn, modified these, it could create substances never before produced by the bacterium. "One or more of these products could be a compound toxic to man."[9] Since L-tryptophan itself is toxic to the bacteria in high concentrations, in an act of self-preservation, the bacteria might have created an enzyme to alter the L-tryptophan. One way or another, something new had started appearing in Showa Denko's product.

Biotech Alarm

The possibility that genetic engineering might be responsible for the deadly EMS epidemic posed quite a threat to the young biotech industry. If consumers linked the new science with the horrid symptoms of the disease, the industry might have to spend years or decades to win back public confidence. At the very least, new regulations might force them to submit products for costly safety testing, something they had thus far been able to avoid.

But the word started to spread. On July 11, 1990, the *Journal of the American Medical Association (JAMA)* published a study mentioning for the first time that Showa Denko's bacteria had been genetically engineered. In fact, the company had introduced a new GM bacterium called Strain V in December 1988, a few months before the epidemic hit.

On August 14, *Newsday* ran a story entitled "Genetic Engineering Flaw Blamed for Toxic Deaths". The article quoted Michael Osterholm, an epidemiologist at the Minnesota Health Department and co-author of a study on EMS published that month in the *New England Journal of Medicine*. He said, "Strain V was cranked up to make more L-tryptophan and something went wrong. This obviously leads to that whole debate about genetic engineering."[10] The *Newsday* article inspired a flood of other papers also to report the EMS story as a genetic engineering problem.

To stem the tide of anti-biotech sentiment, the industry relied on a spokesman it would later count on year after year—the FDA. In an article in *Science* magazine in late August, Sam Page, chief of the natural products and instrumentation branch at the FDA, "blasted Osterholm for 'propagating hysteria'. The whole question: Is there any relation to genetic engineering? is premature—especially given the impact on the industry."[11]

Osterholm countered: "Anyone who looks at the data comes to the same conclusion. . . . I think FDA doesn't want it to be so because of the implications for the agency."

According to the article, the FDA knew for months that the contaminated L-tryptophan was created by GM bacteria, but withheld the information from the public "apparently hoping to keep the recombinant link quiet until they could determine whether it in fact did play a role in the outbreak."[11]

In spite of any damage Osterholm's remark may have had on the biotech industry, the study he helped co-author inadvertently gave GMO proponents an alternative explanation for the epidemic that they continue to use to this day. Shortly after Strain V was introduced, Showa Denko made another change in its L-tryptophan production process. The company reduced the amount of carbon powder in its filters from twenty to ten kilograms per batch.

Carbon filters are used to remove contaminants that are created during the fermentation process. Showa Denko officials claimed that the ten kilos of carbon continued to produce a product that was within the specifications required by US pharmaceutical standards: purity of 98.5 percent or better. Nonetheless, it was possible that the change in the filter could have allowed deadly trace contaminants to get through. Hence the new argument: the culprit was not genetic engineering but a change in the filtration.

This alternative hypothesis appears to have saved the reputation of the biotech industry, allowing GM food and supplements to continue to be sold without safety testing. Let's analyze this alternative hypothesis to see if it justifies the FDA's hands-off approach.

The Blame Game

In the *New England Journal of Medicine* article Osterholm and his colleagues explain, "Although the powdered carbon may have contributed to the removal of the [toxic] agent, it does not explain how the agent was introduced into the product."[12] Showa Denko's own lawyer admits the filter hypothesis "is discounted by scientists at Showa Denko". He said that "the amount of powdered carbon used for filtration had varied before . . . and it was not unusual for it to dip this low."[11]

Strain V was considered the superman of all the previous strains used by the company. Its output was about twice that of Strain I. Osterholm argued in his paper that the newly introduced Strain V bacterium "may have produced larger quantities of the [toxic] agent than earlier strains."[12] Similarly, Yanofsky points out that the higher amount of L-tryptophan in the fermentation process increased the possibility that side reactions would produce more contaminants. He says, "It's possible that one purification scheme may be quite adequate when producing low levels of tryptophan, but at higher levels, it might not be good enough."[13] Thus, as Showa Denko introduced a genetically engineered strain that likely produced more contaminants, they reduced their filtration at a time when an increase may have been needed.

If genetic engineering was responsible for producing the contaminants, it provides an explanation for something else that has baffled researchers. Showa Denko's records reveal that the amount of contamination in the L-tryptophan produced from Strain V varied considerably. The lots produced in March, April and May of 1989, for example, contained very high amounts of contamination. Levels of one contaminant dropped unexpectedly toward the end of April and all suspected contaminants had decreased considerably by the time L-tryptophan was taken off the market.[14] These changes, which have baffled researchers, may have resulted from unstable gene expression caused by insertion mutation, genetic hot spots, or other unpredictable effects of genetic engineering.

The Pre-epidemic Cases Argument

Biotech companies and the FDA offer a second reason why genetic engineering was not the most likely cause of EMS: Some of the cases occurred before Strain V was introduced. According to William Crist, an investigative reporter who has spent years studying the EMS tragedy, the FDA's biotechnology coordinator, James Maryanski, told him in a July 1996 phone conversation, "We can not rule [genetic engineering] out. . . . However, we are aware of close to two dozen cases of L-tryptophan linked EMS that occurred before Showa Denko began using their engineered strain. So, there would have to be a cause other than just the mere engineering of the strains. Now, I can't say that definitively because we don't have a lot of information on these earlier cases." Maryanski asserted that "either L-tryptophan itself, or L-tryptophan in combination with something that was the result of the purification process, was probably the more likely cause."[9]

Crist was not convinced and decided to investigate. He discovered that the actual number of pre-epidemic cases was considerably higher than Maryanski had described. While the CDC identified nearly 100 cases that began several years prior to the May 1989 epidemic,[15] the actual figure is probably between 350 and 700,[9] because the agency's passive surveillance system identified only one out of every four cases—or even less.

To test Maryanski's claim that L-tryptophan by itself may have been responsible, Crist tried to find out if there were any EMS cases associated with a different company's brand. "I faxed and called about a dozen law firms who had handled [hundreds of] Showa Denko cases. None had handled or knew of even one definite case associated with another manufacturer," reported Crist. Stephen Sheller, an attorney whose firm handled over 100 EMS cases, including about ten with onset prior to the epidemic, said, "We have always been suspicious that there were EMS cases caused by other L-tryptophan. . . . However, we have never had a case that we could confirm that with. All cases that we've had [have] been traced to Showa Denko."[9]

In the scientific literature, Crist found three studies by CDC epidemiologists that showed that Showa Denko's product was associated with cases prior to the epidemic. Moreover, no studies anywhere implicated any brand of L-tryptophan other than Showa Denko's.

These findings contradict the FDA's claim that L-tryptophan itself may have caused the epidemic. If it had, according to CDC epidemiologist Edwin Kilbourne, "all tryptophan products of equal dose produced from different companies should have had the same [effect]."[16] But Kilbourne insists there is no evidence to support this. Likewise, Gleich at the Mayo Clinic says, "Tryptophan itself clearly is not the cause of EMS in that individuals who consumed products from companies other than Showa Denko did not develop EMS. The evidence points to Showa Denko product as the culprit and to the contaminants as the cause."[17]

But that still leaves the question, What was responsible for the pre-epidemic cases? According to Maryanski, since EMS cases "occurred before Showa Denko began using their engineered strain . . . there would have to be a cause other than just the mere engineering of the strains."[9] Maryanski was describing Strain V, introduced in December 1988 and subsequently linked to the EMS epidemic of the following year. Crist, however, discovered in his investigation that *previous strains of bacteria used by Showa Denko were also genetically modified*. Between 1984 and 1988, the company introduced four successive GM strains, II–V.

Thus, the pre-epidemic cases of EMS also appear to be the result of L-tryptophan created from genetically engineered bacteria. This would explain why researchers found that only people who consumed Showa Denko's brand contracted EMS. And since Strains II–V were progressively modified to produce greater and greater quantities of L-tryptophan, each improved strain created more contaminants. This would explain the gradual increase of EMS cases leading up to the Strain V-related epidemic.

In the beginning, Crist wondered why the FDA didn't know about the earlier strains. They had access to the same sources of information he had, and certainly much more. But as Crist was reading a key document about the case, he happened to notice the fax imprint at the top of the page, displaying the date and sender of the document. Crist had found a smoking gun of his own. On October 2001, he wrote to Maryanski and Joseph Levitt, director of FDA's Center for Food Safety and Nutrition: "I have a copy of a September 17, 1990 fax from FDA, which appears to be a report by Showa Denko listing the parent mutant Strain I, and the genetic modifications in Strains II–V. So the agency knew that Showa Denko had used three other strains of engineered bacteria, besides the strain (V) that was linked to the epidemic cases, and did not disclose this fact to the public."

Crist continued, "It appears that FDA has tried to defuse and downplay the issue of genetic engineering by shifting the blame to tryptophan itself, using pre-epidemic EMS . . . cases as justification."[9] The FDA did not respond to his letter.

Continuing to pore over data, Crist discovered that earlier strains of Showa Denko's GM bacteria also produced contaminated product. In fact, a German firm had rejected shipments of Showa Denko's L-tryptophan in 1988, prior to Strain V, due to the presence of an impurity. He wrote, "According to internal Showa Denko documents, when Showa Denko was questioned about the . . . impurity, they admitted that they couldn't determine a lack of toxicity of the impurity because they couldn't figure out what the impurity was."[9]

After the outbreak, a study that tracked one of the EMS-related contaminants (Peak E/EBT) identified it in pills manufactured as early as August 19, 1986. Thus, both Strains III and IV produced this impurity. And one person who contracted a severe case of EMS in November 1987 was taking L-tryptophan produced from Strain III. His pills were tested and identified as Showa Denko's with their characteristic pattern of impurities.[9]

Showa Denko also tested their products, but some of their most important test results are no longer available. According to John Baker, an attorney who represented several EMS victims and served as a member of the National Steering Committee for litigation against Showa Denko, "after reviewing the company documents and the depositions of company employees, expert scientists retained by Plaintiffs in the EMS litigation in the United States have opined that Showa Denko appears to have destroyed some of the serial chromatograms that showed contaminants in their L-tryptophan product in 1988."[9]

In what appears to be a tacit acceptance of responsibility, however, Showa Denko did extend out-of-court settlements to pre-epidemic cases of EMS—those who had taken L-tryptophan created from its earlier GM strains of bacteria. In total, the company paid a total of over $2 billion to more than 2,000 victims.

Misdiagnosed Earlier Cases

The number of victims who got EMS from L-tryptophan produced by earlier strains of Showa Denko's GM bacteria may be much higher than originally thought, because identification of the disease can take years. An article in the June 2001 *National EMS Network Newsletter* states, "While many believe there were pre-epidemic cases of EMS that were either given another diagnosis or no diagnosis at all because the term 'eosinophilia-myalgia syndrome' was not a part of the medical lexicon, some people are only now discovering that they have EMS." The way that a pre-epidemic EMS case is determined is "first by establishing that a person in fact consumed L-tryptophan produced by Showa Denko (SDK)."[18] Studies show that many patients diagnosed with eosinophilic fasciitis (EF), fibromyalgia and scleroderma had taken L-tryptophan and therefore may have had EMS but were misdiagnosed.

Don Hudson of Louisiana is one such patient. He says, "My first EMS symptoms began in November 1987. By February 1988, I was extremely ill and had been hospitalized. My treating physician was totally baffled. He ran every test conceivable, including a muscle biopsy. I was on the edge of death and slipping fast. For whatever reason, I stopped taking L-tryptophan, and within a month my condition had improved past the critical stage. My doctor diagnosed the problem as fibromyalgia, but he told us that my illness was one of those things that medicine just couldn't explain."

When Hudson attended a fibromyalgia support group, he quickly became aware that his situation was quite different from the others. "None of the other members had life-threatening symptoms. Practically all gave me a blank look when I asked how high their eosinophils had gotten. I had an eosinophils count of 25,000 to 58,000, whereas zero to 400 is normal."[9] Hudson's count remains high and, like many other EMS victims, he continues to face symptoms every day. Hudson battles temporary blindness, irritable bowel, unbearable muscle pain, fatigue, tremors and breathing trouble, among other problems.

Someone Was Not Cooperating

When Strain V was first implicated, FDA investigators should naturally have obtained the bacterium to verify that it produced the contaminated L-tryptophan. But Maryanski told Crist in an interview that the FDA never obtained samples of the bacterial strain. An article in *Science* claims that Showa Denko destroyed all of their bacteria when the toxicity problems first emerged.[11] But when Crist contacted Showa Denko's attorney Don Morgan in March 2001, he heard quite a different story. According to Crist, Morgan revealed "that Showa Denko offered to give FDA the cultures, but they did not want to mail them, as apparently FDA had requested." When exposed to the environment, the bacteria can mutate and produce additional impurities. Morgan told Crist "that FDA never followed through on Showa

Denko's offer to turn over the bacteria, in such a way that they could show FDA the proper way to handle them, etc. [Morgan said] the company finally destroyed the bacteria in 1996."[9]

Crist wrote to Sam Page, then a scientific director at the FDA, asking him to respond to Morgan's claim. His questions remained unanswered. Crist submitted several Freedom of Information (FOI) requests to the FDA and CDC in 1998 and again in 2001. Crist said, "In 1998, responses were received from FDA and CDC, but none of the specific documents and/or information requested were supplied. In 2001, FDA FOI staff said that the information requested either 'was lost' or 'could not be found', and that the people who were involved at that time (1989–90) had all left FDA." Crist told the FDA representative that several of the scientists and officials to whom his requests were directed, including Sam Page, Rossanne Philen, Henry Falk and Edwin Kilbourne are all still at the FDA or CDC, but the FDA FOI staff person repeated that the people involved had left. Crist says, "Their failure to respond suggests that the questions may, in fact, be on target. . . . Now, it appears that they both may have known all along that the GE strains did play a crucial role in EMS and that they concealed this information to protect the US biotech industry."[9]

The FDA Takes the Stand
On July 18, 1991, Douglas Archer, deputy director of the FDA's Center for Food Safety and Applied Nutrition, sat before a congressional committee to give the agency's official version of the EMS incident. Observers familiar with the FDA's pro-biotech bias were waiting to see how Archer was going to present this sensitive issue of genetically engineered L-tryptophan to the lawmakers. They knew the FDA was determined to maintain control of GMO policy, and that the agency did not want Congress to step in to draft new laws. Rather, the FDA was developing its own pro-industry policy based on food laws written prior to genetic engineering. The observers waited . . . and waited . . . and waited. Nothing. There was not a single mention of genetic engineering in Archer's testimony.

Certainly Archer knew about the genetically engineered bacteria.* But instead of risking public condemnation of genetic engineering and con-

---

* In fact, in 2001, when I mentioned to a former FDA employee that the FDA representative speaking before Congress about L-tryptophan failed to mention that it was genetically modified, she said she found that hard to believe. "Everyone in the agency knew it was genetically altered," she said.

gressional meddling, Archer used the opportunity to advance the FDA's own agenda. In his overview, he stated: "The L-tryptophan/ EMS incident that is the focus of this hearing, and especially the suffering and death of those who used the products in question, demonstrate the dangers inherent in the various health fraud schemes that are being perpetrated upon segments of the American public."

Why did the FDA choose to target "health fraud schemes"? Archer said in his testimony that "there was an agency desire to closely regulate the addition of all vitamin, mineral and amino acids used in foods, including dietary supplements." By making L-tryptophan the enemy, the FDA was able to use it as a scapegoat to justify that food supplements were dangerous and needed regulation. In fact, the FDA had already tried to stop L-tryptophan from being sold over-the-counter.

Congress had not cooperated with the FDA's desire to regulate and restrict supplements. In 1976, Congress passed the Proxmire Amendment, which prevented the FDA from putting a limit on the potency of a vitamin or mineral supplement based on what the agency "considered to be rational or useful". In his testimony, Archer seemed to take Congress to task for limiting FDA control over supplements. "The so-called Proxmire Amendment to the FDC Act is another factor influencing the atmosphere in which the agency made its decisions regarding the regulation of amino acids," he said. "The amendment was passed in direct response to an FDA rulemaking effort and it seemed to signal Congressional intent that supplement-type products not be regulated without indication of real danger to health."

Archer promoted the agency's get-tough-on-non-drugs policy. He continued, "As part of Commissioner Kessler's announced program of stepped-up enforcement, he has stated unequivocally that the FDA will not tolerate unsubstantiated drug claims being made for foods, including amino acid supplements." He added, "Another aspect of this problem is that some segments of the supplement industry have been able to capitalize on the confusion created by health claims for foods." Archer also admonished doctors, telling them not to recommend L-tryptophan to their patients for drug purposes.

Archer drew a distinction between doctor-prescribed L-tryptophan used in infant formula, intravenous administration and feeding tubes, which the agency had deemed legal, and the over-the-counter variety, which they had not approved. Twice they had tried to remove what Archer called the illegally marketed variety by bringing sellers to court. Both times they lost. In 1990, when EMS was linked to L-tryptophan consumption, the FDA finally got its way and banned the over-the-counter variety. Thus, the EMS epidemic had helped them accomplish what two FDA lawsuits were been unable to do. "Finally, on February 19, 1991," Archer said, "because of the strong epidemiological relationship between EMS and L-tryptophan manufactured by [Showa Denko], FDA expanded the recall to include legal products containing Showa Denko L-tryptophan."[7]

In the end, Archer's carefully worded testimony accomplished a great deal for the FDA. It affirmed the agency's need for more freedom to regulate supplements and even praised the FDA for having tried to take this harmful supplement off the shelves for years. Further, because Archer did not mention genetic engineering, Congress made no inquiries on the subject and the media also avoided it, blaming the epidemic instead on unregulated health schemes. L-tryptophan remains off the market, except if prescribed by a physician.

Current Regulations Would Approve the L-Tryptophan

If introduced today for the first time, the contaminated L-tryptophan could pass easily through the current FDA regulations. Even the presence of the impurities in Showa Denko's L-tryptophan would not have stopped distribution because they were not known toxins. The FDA "can detect the presence of known toxins based only on known properties of preexisting food."[5]

There are many supplements manufactured today that use GM bacteria. A form of GMO-derived vitamin B-2, for example, was approved in the UK on the basis of data that identified only those contaminants found at levels above 0.1 percent. The impurities in L-tryptophan, however, were ten times smaller.[19] According to Stephen Naylor of the Mayo Clinic, "the presence of the contaminants in the [Showa Denko] L-tryptophan is astonishingly small and so you require very sophisticated instrumentation and a lot of hard work to even come close to determining the structures."[9] Showa Denko monitored the levels of impurities in their L-tryptophan daily. It remained within US standards.

Similarly, according to the BBC magazine, the deadly batches of tryptophan would likewise have been approved in Europe. "It would only be when people began dropping like flies that the problem would become apparent."[20] It's sobering to consider that it was the fact that people *did* drop like flies that helped catch the disease. Crist and others compare L-tryptophan to the fertility drug thalidomide, which was responsible for severe birth defects in the late 1950s and early 1960s: "We emphasize that if thalidomide had happened to cause a type of birth defect that was already common, e.g., cleft palate or severe mental retardation, we would still not know about the harm, and pregnant women would have kept on taking it for its undoubted benefits. The fractional addition to [birth defect] figures that were already relatively large would not have been *statistically* significant. But as it turned out, the damage noticed was of a kind that most doctors never see in a whole career—drastic malformations of the arms and legs—so although the numbers were not huge, these cases were picked up."

Crist points out that it was the uniqueness of EMS that allowed the L-tryptophan problem to surface. If, on the other hand, Showa Denko's contaminated supplements "had caused the same numbers of a common illness instead, say asthma, we would still not know about it. Or if it had caused delayed harm, such as cancer twenty to thirty years later, or senile dementia in some whose mothers had taken it early in pregnancy, there would have been no way to attribute the harm to the cause."[21]

In spite of the severity of the outbreak, it still took years to trace the crippling disease to L-tryptophan and months more to discover that Showa Denko's brand was responsible. One contributing factor, of course, was that there was no label to distinguish the GM version from the natural variety. The same holds true for GM foods. "In the absence of labelling for genetically modified products," write Rampton and Stauber in *Trust Us We're Experts*, "it is impossible to determine who has been eating mutant soybeans and who has been eating natural ones. If something toxic enters the food supply, tracing it to its source will be difficult if not impossible."[5]

Wisdom of Squirrels, Elk, Deer, Raccoons and Mice

For years, a retired Iowa farmer fed squirrels on his farm through the winter months by placing corncobs on feeders. One year, just for the heck of it, he decided to see if the squirrels had a preference for Bt maize or natural maize. He put natural maize in one feeder and Bt maize in another about twenty feet away. The squirrels ate all the maize off the natural cobs but didn't touch the Bt. The farmer dutifully refilled the feeder with more natural maize and sure enough, it was soon gone. The Bt, however, remained untouched.

The retired farmer got curious. What if the Bt variety was the squirrels' only choice? To find out, he didn't refill the natural maize. At the time, Iowa was plunged into the coldest days of winter. But day after day, the Bt cob remained intact. The squirrels went elsewhere for their food. After about ten days, the squirrels ate about an inch off the tip of an ear, but that's all. The farmer felt sorry for the squirrels and put natural maize back into the feeders, which the squirrels once again consumed.[1]

"A captive elk escaped and took up residence in our crops of organic corn [maize] and soy. It had total access to the neighbouring fields of GM crops, but never went into them."[2]
—Susan and Mark Fitzgerald, Minnesota

Writer Steve Sprinkel described a herd of about forty deer that ate from the field of organic soybeans, but not the Roundup Ready variety across the road. Likewise, raccoons devoured organic maize, but didn't touch an ear of Bt maize growing down the road. "Even the mice will move on down the line if given an alternative to these 'crops.'"[3]

A farmer in Holland verified the food preference of mice when he left two piles of maize in his mice-infested barn. One pile was genetically modified; the other was natural. The GM pile was untouched while the non-GM pile was completely eaten up.[4]

Chapter 5

Government By the Industry, For the Industry

Vice President George Bush sat in his chair across from four Monsanto executives. They had come to the White House with an unusual request. They wanted *more* regulation. They were venturing into a new technology, the genetic modification of food, and they were actually asking the government to oversee their emerging industry.

But this was late 1986. Ronald Reagan was president and the administration was busily *deregulating* business. Bush needed convincing. "We bugged him for regulation," said Leonard Guarraia, one of the executives at the meeting. "We told him that we have to be regulated."[1]

Monsanto was about to make a multibillion-dollar gamble. With this new technology, they could engineer and patent a whole new kind of food. Later, by buying up seed companies around the world, Monsanto could replace the natural seeds with their patented engineered seeds and control a hefty portion of the food supply.

But there was fear among Monsanto's ranks—fear of consumers' and environmentalists' reactions. Their fear was borne of experience. Years earlier, Monsanto had assured the public that their Agent Orange, the defoliant used during the Vietnam War, was safe for humans. It wasn't. Thousands of veterans and tens of thousand of Vietnamese who suffered a wide range of maladies, including cancer, neurological disorders and birth defects, blame Monsanto.

Monsanto had also declared their electrical insulator poly-chlorinated biphenyls (PCBs) as safe. They weren't. Outlawed in 1978, they have been linked to cancer and birth defects, and are considered a major environmental hazard. According to court documents, Monsanto executives knew that its PCB factory in Anniston, Alabama was exposing the town's population to serious health risks. They regularly dumped PCBs in the town but covered up the fact for more than forty years. According to the

Washington Post, "In 1966, for example, Monsanto managers discovered that fish dunked in a local creek turned belly-up within ten seconds, spurting blood and shedding skin as if dropped into boiling water. In 1969, they found a fish in another creek with 7,500 times the legal PCB level. But they never told their neighbours, and concluded that 'there is little object in going to expensive extremes in limiting discharges.'" One internal memo stated, "We can't afford to lose one dollar of business."[2]

On February 22, 2002, Monsanto was found guilty of negligence, wantonness, suppression of the truth, nuisance, trespass and outrage. "Under Alabama law," the *Washington Post* article explained, "the rare claim of outrage typically requires conduct 'so outrageous in character and extreme in degree as to go beyond all possible bounds of decency so as to be regarded as atrocious and utterly intolerable in civilized society.'"[2]

The public's reprisal to these and other Monsanto mistakes was considerable. A former Monsanto vice president admitted, "We were despised by our customers."[3] With genetic engineering, Monsanto knew they needed a new approach. They were determined to work with potential critics in advance to win over their support. "Active outreach," according to the strategy committee's plan, "will encourage public interest, consumer and environmental groups to develop supportive positions on biotechnology."[1]

Their plan, dated October 13, 1986 and subsequently obtained by the *New York Times,* also prescribed that Monsanto engage with regulators and elected officials worldwide, create "support for biotechnology at the highest US policy levels", and gain endorsements in both the Democratic and Republican party's 1988 presidential platforms.

They also needed federal regulations. With that in place, it would be the government, not Monsanto, who would be assuring the public that GM products were safe. Monsanto wasn't ready to once again ask the public to "trust us".

Monsanto's connections in Washington ran deep—very deep. The meeting with Vice President Bush was successful, and they got what they asked for. According to the *New York Times*, "It was an outcome that would be repeated, again and again, through three administrations. What Monsanto wished for from Washington, Monsanto—and, by extension, the biotechnology industry—got."

But in the early 1990s, the president of Monsanto, who favoured the cautious, collaborative approach, retired. The task of overseeing the expansion of genetically engineered food was given to the enthusiastic Robert Shapiro. He "shelved the go-slow strategy of consultation and review" and brought the GM campaign up to ramming speed. According to the *New York Times*, "Monsanto would now use its influence in Washington to push through a new approach." To help Monsanto "speed its foods to market, the White House quickly ushered through an unusually generous policy of self-policing."[1]

Monsanto's influence was legendary. Washington insiders watched with astonishment as the company dictated policy to the Agriculture Department (USDA), Environmental Protection Agency (EPA) and ultimately the Food and Drug Administration (FDA).

According to Henry Miller, who was in charge of biotechnology issues at the FDA from 1979 to 1994, "the US government agencies have done exactly what big agribusiness has asked them to do and told them to do."[1]

This biotech industry worked its magic with the Council on Competitiveness, a senior policy-making group created by President Bush in March 1989. Vice President Dan Quayle was put in charge, "with responsibility for reducing the regulatory burden on the economy." The council was also assigned to counter the drastic US trade deficit by making American goods more competitive in overseas markets. Members of this elite council included "The Attorney General, Secretary of Commerce, Director of [the Office of Management and Budget] and Chair of the Council of Economic Advisors. . . . The President's chief of staff coordinated Council activities."[4]

The biotech industry's success with these government leaders became apparent on May 26, 1992 in the Indian Treaty Room of the Old Executive Building. There, Vice President Dan Quayle announced the Bush administration's new policy on genetically engineered food: "The reforms we announce today will speed up and simplify the process of bringing better agricultural products, developed through biotech, to consumers, food processors and farmers. We will ensure that biotech products will receive the same oversight as other products, instead of being hampered by unnecessary regulation."[1]

By "receive the same oversight as other products", Quayle meant that GM foods would be considered just as safe as natural, non-GM foods. And sidestepping "unnecessary regulation" meant that the government would not require any safety tests or any special labels identifying the foods as genetically engineered. The rationale for this hands-off policy was spelled out in an FDA document dated three days after Quayle's announcement. "The agency is not aware of any information showing that foods derived by these new methods differ from other foods in any meaningful or uniform way."[5]

Monsanto, under its new leadership, had what it wanted: government endorsement of safety and no regulations that would interfere with its plans for rapid worldwide sales.

Political Science at the FDA

Attorney Michael Taylor had overseen the development of FDA policy. Prior to working at the agency, Taylor worked at King and Spaulding law firm; Monsanto was his personal client. Taylor helped Monsanto draft pro-biotech regulations that the industry would lobby for. While working for the FDA, Taylor could implement those laws himself. For Monsanto, there was no better person to step into a leadership role at the FDA.

Taylor did not simply fill a vacant position at the agency. In 1991 the FDA created a new position for him: Deputy Commissioner for Policy. He instantly became the FDA official with the greatest influence on GM food regulation, overseeing the development of government policy.

According to public interest attorney Steven Druker, who has studied the FDA's internal files, "During Mr. Taylor's tenure as Deputy Commissioner, references to the unintended negative effects of bioengineering were progressively deleted from drafts of the policy statement (over the protests of agency scientists), and a final statement was issued claiming (a) that [GM] foods are no riskier than others and (b) that the agency has no information to the contrary."[6] In 1994, Taylor became the administrator at the Department of Agriculture's Food Safety and Inspection Service, where he was also involved in biotechnology issues. He later became Vice President for Public Policy at Monsanto.

When the FDA announced its policy, the public was not aware of any internal dissent. The policy boldly claimed that there was no information to indicate that GM foods were different or more risky than natural varieties. Since the American public generally trusts the FDA, they assumed that no such risks existed. But nearly a decade later, the agency's internal documents—made public for the first time through a lawsuit—told a different story.

Linda Kahl, an FDA compliance officer, protested that by "trying to force an ultimate conclusion that there is no difference between foods modified by genetic engineering and foods modified by traditional breeding practices", the agency was "trying to fit a square peg into a round hole". She insisted, "the processes of genetic engineering and traditional breeding are different, and according to the technical experts in the agency, they lead to different risks."[7]

One such expert was FDA microbiologist Louis Pribyl. "There is a profound difference between the types of unexpected effects from traditional breeding and genetic engineering," wrote Pribyl in a letter to James Maryanski, the FDA's biotech coordinator. Pribyl said that several aspects of gene splicing "may be more hazardous".[8] According to the *New York Times*, "Dr. Pribyl knew from studies that toxins could be unintentionally created when new genes were introduced into a plant's cells."[1] Moreover, Pribyl wrote "there is no certainty that [the breeders of GM foods] will be able to pick up effects that might not be obvious." He declared, "This is the industry's pet idea, namely that there are no unintended effects that will raise the FDA's level of concern. But time and time again, there is no data to back up their contention."[8]

Pribyl was one of many FDA scientists asked to provide input during the formulation of the FDA's policy on genetically engineered food. According to Druker, records show that the majority of these scientists identified potential risks of GM foods. Druker was the main organizer of the lawsuit that forced the FDA documents into the public domain; his non-profit organization, the Alliance for Bio-Integrity, was the lead plaintiff. Having sorted through tens of thousands of pages of FDA documents, he described the opinion of the agency's scientists as follows: "The predominant view was that genetic engineering entails distinct risks and that its products cannot be regarded as safe unless they have been confirmed to be so through appropriate feeding studies." Druker says several scientists "issued strong warnings".[6]

The Toxicology Group, for example, warned that genetically modified plants could "contain unexpected high concentrations of plant toxicants", and described the reasons why these might be very difficult to identify.[9] Their director wrote, "The possibility of unexpected, accidental changes in genetically engineered plants justifies a limited traditional toxicological study."[10]

The Division of Food Chemistry and Technology outlined four potential dangers:

1. "Increased levels of known naturally occurring toxins",
2. "Appearance of new, not previously identified" toxins,
3. Increased tendency to gather "toxic substances from the environment" such as "pesticides or heavy metals", and
4. "Undesirable alterations in the levels of nutrients".

They warned that "unless genetically engineered plants are evaluated specifically for these changes", these four "may escape breeders' attention". The division recommended testing every GM food "before it enters the marketplace".[11]

Gerald Guest, the director of FDA's Center for Veterinary Medicine (CVM) sent a letter to Maryanski saying that he and the other CVM scientists concluded that there is "ample scientific justification" to require testing and review of each GM food before it is eaten by the public. He stated, "CVM believes that animal feeds derived from genetically modified plants present unique animal and food safety concerns." He pointed out that, "residues of plant constituents or toxicants in meat and milk products may pose human food safety concerns."[12] Guest also wrote, "I would urge you to eliminate statements that suggest that the lack of information can be used as evidence for no regulatory concern."[12]

In spite of repeated internal memos outlining the potential for increased health risks posed by this new technology, subsequent drafts of the FDA's policy statement, overseen by Taylor, deleted more and more of the scientist's input. In a fiery memo to Maryanski, Pribyl challenged the direction the policy statement had taken: "What has happened to the scientific elements of this document? Without a sound scientific base to rest on, this becomes a broad, general, 'What do I have to do to avoid trouble'-type document. . . . It will look like and probably be just a political document. . . . It reads very pro-industry, especially in the area of unintended effects, but contains very little input from consumers and only a few answers for their concerns."

Pribyl pointed out a glaring inconsistency. He said that while the FDA policy says "that there are no differences between traditional breeding and [genetic modification] . . . In fact the FDA is making a distinction, so why pretend otherwise." Pribyl also made two eerily accurate predictions:

1. "Industry will do what it HAS to do to satisfy the FDA 'requirements' and not do the tests that they would normally do because they are not on the FDA's list;" and

2. "There will be . . . less concern about safety, because of a false sense of 'knowing what one is doing' and 'it's been done hundreds of times before without a problem, why check it now.'"[8]

But while the FDA's scientists were emphasizing caution and testing, its leaders were beholden to an altogether different lobbying effort. A March 1992 memo from FDA Commissioner David Kessler confirmed the White House's influence in the crafting of the agency's policy. "The approach and provisions of the policy statement are consistent with the general biotechnology policy established by the Office of the President. . . . It also responds to White House interest in assuring the safe, speedy development of the US biotechnology industry."[13]

But even the draft of the policy that Kessler praised as White House-friendly was subject to further revision as it went up the political chain of command. A May 1992 Memorandum from the Office of Management and Budget to President Bush's White House counsel made the following recommendations. "The policy statement needs to stress the role of decentralized safety reviews by producers; with informal FDA consultation only if significant safety or nutritional concerns arise. It should avoid emphasizing obligatory FDA review and oversight." The letter also suggested that the following sentence about genetic engineering be added. "Since these techniques are more precise, they increase the potential for safe, better characterized and more predictable foods."[14]

Similarly, a memo from the Office of the Assistant Secretary for Health, at the Department of Health & Human Services, expressed reservations about the length and depth of the policy statement's concern for environmental effects of GM crops. The letter said, "The extensive twelve-page discussion seems to be . . . dangerously detailed and drawn-out. . . . In contrast to the sections on food safety, which properly imply that biotechnology is a fundamentally innocuous tool of food production and that the fruits of biotechnology will be substantially equivalent to those with which we are already familiar, the [environmental] section gives an incorrect impression that biotechnology raises significant new agricultural and environmental concerns."[15]

These memos reveal that as the evaluators have less and less background in science and more political accountability, the foods, and their environmental impact, are regarded as safer and safer. In the end, it was the political, rather than scientific recommendations, that prevailed. The agency not only ignored its scientists, it claimed their concerns never existed. The official FDA policy proclaiming ignorance of any meaningful differences between GM and non-GM food became the rationale for eliminating any meaningful oversight. Other government departments also invoked this political concept of equivalence in support of their policies. For example, the State Department's Melinda Kimble, while negotiating GMO trade policy, said, "I want to make very clear that it is the position of the United States government that we do not believe there is a difference between GMO commodities and non-GMO commodities."[16] Likewise, a March 2003 statement by Speaker of the House Hastert declared, "There is general consensus among the scientific community that genetically modified food is no different from conventional food."[17]

When the FDA documents eventually became public, Maryanski defended the agency's policy. On February 28, 2000, he told the OECD Conference on GM Food Safety in Edinburgh that the FDA scientists had merely been asking questions about the various issues involved in bioengineered food. Maryanski was unpleasantly surprised when Druker, who was a member of the conference, stood up and invited the audience to read the FDA memos that were posted on his organization's website. They could see for themselves that the agency's scientists were not merely asking questions; many of their statements were quite emphatic about the unique risks of GM foods.

Maryanski, other FDA officials and representatives throughout the US government continue to claim that there is overwhelming consensus among scientists that GM foods are safe. In an October 1991 letter to a Canadian official, however, Maryanski himself had admitted that this was not true. He said, "there are a number of specific issues . . . for which a scientific consensus does not exist currently, especially the need for specific toxicology tests." Maryanski also said, "I think the question of the potential for some substances to cause allergenic reactions is particularly difficult to predict."[18]

Commenting on statements made by FDA scientists, the *New York Times* wrote: "The scientists were displaying precisely the concerns that Monsanto executives from the 1980s had anticipated — and indeed had considered reasonable. But now, rather than trying to address those concerns, Monsanto, the industry and official Washington were dismissing them as the insignificant worries of the uninformed."[1]

Many scientists who understood the dangers, however, were not convinced by the FDA's assurances. Geneticist David Suzuki, for example, said, "Any politician or scientist who tells you these products are safe is either very stupid or lying. The experiments have simply not been done."[19] A January 2001 report from an expert panel of the Royal Society of Canada likewise supported the conclusions of the FDA scientists. The report said it was "scientifically unjustifiable" to presume that GM foods are safe. The report explains that the "default prediction" for any GM foods is that "expression of a new gene (and its products) . . . will be accompanied by a range of collateral changes in expression of other genes, changes in the pattern of proteins produced and/or changes in metabolic activities." This could result in novel toxins or other harmful substances. The report emphasized the need for safety testing, looking for short- and long-term human toxicity, allergenicity and other health effects.[20] The panel began their comprehensive 245-page report by quoting the editors of the UK's *Nature Biotechnology*. "The risks in biotechnology are undeniable, and they stem from the unknowable in science and commerce. It is prudent to recognize and address those risks, not compound them by overly optimistic or foolhardy behaviour."[20]

Rotten Tomatoes

While the FDA was busily crafting their industry-friendly GMO policy in the early 1990s, Calgene was preparing to introduce the world's first genetically modified food crop: the FlavrSavr Tomato. Gifted with mythical endurance, this GM wonder could remain looking fresh for weeks after being picked.

Although the FDA did not require it, Calgene voluntarily did three feeding studies with rats and sent the results to the FDA for its blessing. Internal FDA documents show that the agency scientists were concerned about the presence of stomach lesions. Among the female rats in one study, seven of the forty rats that ate the FlavrSavr had lesions; none were found in the controls that ate natural tomatoes.

FDA reviewers repeatedly asked Calgene to provide additional data in order to resolve what they regarded as outstanding safety questions. The director of the FDA's Office of Special Research Skills, wrote: ". . . the data fall short of 'a demonstration of safety' or of a 'demonstration of reasonable certainty of no harm' which is the standard we typically apply to food additives. To do that we would need, in my opinion, a study that resolves the safety question raised by the current data."[21] The Additives Evaluation

Branch agreed that "unresolved questions still remain",[22] and the staff pathologist stated, "In the absence of adequate explanations by Calgene, the issues raised by the Pathology Branch . . . remain and leave doubts as to the validity of any scientific conclusion(s) which may be drawn from the studies' findings."[23]

Pusztai, who looked at the study years later, disagreed with the Calgene's conclusions that the lesions "were considered to be of no importance", since, he said, "in humans they could lead to life-endangering haemorrhage, particularly in the elderly who use aspirin to prevent thrombosis."[24] He was similarly amazed that no follow-up examination of the intestines was conducted to see if they were similarly affected. On top of this, Pusztai pointed out that there was no explanation provided as to why another seven of the forty GM-fed rats died within two weeks.

While one group of FDA scientists was assessing the FlavrSavr rat study, another group was asked to evaluate Calgene's proposed use of an Antibiotic Resistant Marker (ARM) gene. As you may recall from Chapter Two, after the cells are blasted with the gene gun they are doused with antibiotics. If the cells survive, it means that the foreign gene made it into the cells' DNA. Calgene wanted to use an ARM gene that would cause its tomato cells to survive the antibiotic kanamycin.

On December 3, 1992, the Division of Anti-Infective Drug Products submitted to the FDA Biotechnology Coordinator their response to Calgene's proposal with the key sentence of their conclusion emphasized in all capital letters: "IT WOULD BE A SERIOUS HEALTH HAZARD TO INTRODUCE A GENE THAT CODES FOR ANTI-BIOTIC RESISTANCE INTO THE NORMAL FLORA OF THE GENERAL POPULATION."[25]

No ambiguity there.

To further emphasize his concern, the division's director sent the document two weeks later to another FDA official with a cover letter entitled, "The tomatoes that will eat Akron". He added, "You really need to read this consult. The Division comes down fairly squarely against the [kanamycin] gene marker in the genetically engineered tomatoes. I know this could have serious ramifications."[25]

ARM genes are not the only method to confirm that a foreign gene has successfully made it into the DNA. But it's the easiest way. This wasn't a good enough reason for Albert Sheldon, an FDA microbiologist who wrote "Other markers . . . are available and should be used." In a March 1993 memo to Maryanski, Sheldon said, "In my opinion, the benefit to be gained by the use of the kanamycin resistance marker in transgenic plants is out-weighed by the risk. . . . If we allow this proposal, we . . . will probably assure dissemination of kanamycin resistance."[26]

These FDA scientists were aware of the very serious threat posed by infections that resist antibiotics. According to the FDA website, such infections "increase risk of death, and are often associated with prolonged hospital stays, and sometimes complications. These might necessitate removing part of a ravaged lung, or replacing a damaged heart valve."[27] The number of sicknesses and deaths due to resistant infections continues to rise, due in part to the over prescription of antibiotics. According to *BBC Online*, "pessimistic experts believe it is only a matter of time at current rates until virtually every weapon in the pharmaceutical arsenal is nullified."[28] No wonder the FDA division director referred to the FlavrSavr as "The tomatoes that will eat Akron".

In spite of the concerns about antibiotic resistance and the unresolved questions about the feeding trial, the FDA approved the FlavrSavr tomato on May 18, 1994. According to Druker, the FDA "claimed that all relevant safety issues had been satisfactorily resolved and said that because the FlavrSavr had performed so well, it would be unnecessary for any subsequent bioengineered food to be subjected to the same rigorous standard of testing. To date, there is no reliable evidence showing that any has successfully met the standard the FlavrSavr failed to meet."[6]

Druker also points to a statement in one FDA scientist's memo that shows the agency administrators had instructed their scientists to subject GM foods to a lower safety standard than that normally applied to food additives: "It has been made clear to us that this present submission [FlavrSavr rat study] is not a food additive petition and the safety standard is not the food additive safety standard. It is less than that but I am not sure how much less."[21] Druker says that this preferential treatment violates the FDA's own regulations, which state that tests on new foods (such as those produced through genetic engineering) "require the same quantity and quality of scientific evidence as is required to obtain approval of the substance as a food additive."[29]

Mixed Agendas, Problems with Personnel
The approval of GM foods is better appreciated in light of the perennial challenges faced by the FDA. The agency regulates 35 percent of the gross national product, but its budget and resources are tiny by comparison. It

is severely understaffed and has had difficulty attracting and keeping qualified scientists from academia and industry, where more prestige and
higher salaries are available.

James Turner, a long-time FDA watchdog and best-selling author of
*The Chemical Feast: The Nader Report on the Food and Drug
Administration*, describes a three-tiered structure among the personnel at
the agency. "At the top there are political appointees not necessarily bound
by science, but often influenced by other agendas. Many pass through the
agency at a rapid clip, moving from one regulated enterprise to another. At
the same time, some of the finest scientists and public servants that I have
ever met worked at the FDA. Unfortunately many of them are often hired
away by universities, non-profit groups and other public health agencies,
leaving a less dedicated and less competent residue of people not highly
sought after outside of government."[30]

It may have been this third level that Richard Crout, former director of
the FDA's Bureau of Drugs, described in his April 1976 testimony before
the Panel of New Drug Regulations. "I want to describe the agency as I saw
it. No one knew where anything was. . . . There was absenteeism; there was
open drunkenness by several employees, which went on for months; there
was intimidation internally. . . . People—I'm talking about division directors and their staffs—would engage in a kind of behaviour that invited
insubordination; people tittered in the corners, throwing spitballs—now
I'm describing physicians; people would slouch down in their chairs and
not respond to questions; and moan-and-groan, the sleeping gestures. This
was a kind of behaviour I have not seen in any other institution from a
grown man . . . FDA has a long-term problem with the recruitment of personnel, good, scientific personnel."[31]

When President Reagan came into office in 1980 and began his assault
on regulation, the situation at the agency deteriorated even further. With
orders to deregulate, the White House gave the Office of Management and
Budget (OMB) the power to make sweeping changes in all federal agencies. The already understaffed FDA was hit hard.

The influence of the OMB was brought to light in late 1990 when, in
response to the FDA's long delay in establishing some new rules for health
claims on food (unrelated to GM foods), the Congress mounted an investigation of the agency. According to the book *Food Politics* by Marion
Nestle, "The committee concluded that White House interference had
held up the rules for three years and that the FDA's political leaders had

'kowtowed' to the Office of Management and Budget at virtually every step in the process, with the result that the agency's 'regulatory powers [had] been neutered.'"[32]

In 1991, a congressional aide said, "The result of OMB interference [over the previous decade] is that the expertise of scientists and career civil servants is being second-guessed by people who have no legal or scientific basis. . . . At FDA morale stinks. Hundreds of people have either retired or quit in disgust. All the best people, who believed in working on behalf of public health, have gone."[33]

FDA veterinarian Richard Burroughs described the changes he saw. "There seemed to be a trend in the place toward approval at any price. It went from a university-like setting where there was independent scientific review to an atmosphere of 'approve, approve, approve". He said, "The thinking is, 'How many things can we approve this year?' Somewhere along the way they abdicated their responsibility to the public welfare."[33]

This change may have contributed in part to the disturbing finding by the GAO that more than half of the drugs approved by the FDA between 1976 and 1985 had severe or fatal side effects that had not been detected during the agency's review and testing.[34] Thus, after drug companies spent an estimated twelve years and $231 million dollars[35] to research, test and secure new drug approval through a very hands-on FDA approach, more than half of the drugs had to be taken off the market or required major label changes due to missed safety issues.

The Mysterious Changing Hot Potato

The FDA isn't the only government agency that regulates or promotes GM food. The Environmental Protection Agency (EPA) also plays a role, as was illustrated in an October, 1998 article in the *New York Times Sunday Magazine*. The article describes Monsanto's "New Leaf" potato, which creates its own Bt pesticide, and reveals how the EPA and FDA juggle its regulation to satisfy industry desires. The author of the article, Michael Pollan, "was mystified by the fact that the Bt toxin was not being treated as a 'food additive' subject to labelling". The Bt protein was a new ingredient inside the potato being consumed by the public. According to the law any new additive must "be thoroughly tested and if it changes the product in any way, must be labelled". Pollan tells us how he asked the FDA's James Maryanski why the FDA didn't consider Bt a new food additive.

"'That's easy,' Maryanski said. 'Bt is a pesticide, so it's exempt' from

FDA regulation. That is, even though a Bt potato is plainly a food, for the purposes of federal regulation it is not a food but a pesticide and therefore falls under the jurisdiction of the EPA.'"

Pollan asked Maryanski if the safety standards of the EPA are the same as the FDA. "Not exactly," Maryanski said. He explained that while the FDA requires "a reasonable certainty of no harm" in a food additive, pesticides could not meet this standard since, "pesticides are toxic to something. . . . The EPA instead establishes human 'tolerances' for each chemical and then subjects it to a risk-benefit analysis."

When Pollan called the EPA to ask if they had tested Bt potatoes for human safety, their answer was "not exactly". According to Pollan, "The EPA works from the assumption that if the original potato is safe and the Bt protein added to it is safe, then the whole New Leaf package [Bt potato] is presumed to be safe." The EPA figured that the original potato was safe and didn't need testing. They fed Bt toxin to mice and they "did fine, had no side effects".

"In this case there was a small catch," Pollan continues. "The mice weren't actually eating the potatoes, not even an extract from the potatoes, but rather straight Bt produced in a bacterial culture."[36] According to *New Scientist* magazine, "the protein made by the bacteria may not be the same as that made by the plant, especially in its potential to cause allergy."[37] Similarly, most of the twenty-one potential dangers of genetic engineering described in an earlier chapter would have escaped detection using the EPA's testing method.

Pollan looked at a bottle of Bt pesticide used for gardening. Its label warns against "inhaling the spray or getting it in an open wound". He wondered, "If my New Leaf potatoes contain an EPA-registered pesticide, why don't they carry some such label? Maryanski had the answer. At least for the purposes of labelling, my New Leafs have morphed yet again, back into a food: the Food, Drug and Cosmetic Act gives the FDA sole jurisdiction over the labelling of plant foods, and the FDA has ruled that biotech foods need be labelled only if they contain known allergens or have otherwise been 'materially' changed."

"But isn't turning a potato into a pesticide a material change?" asked Pollan.

"It doesn't matter. The Food, Drug and Cosmetic Act specifically bars the FDA from including any information about pesticides on its food labels," was the response.

In addition to the mysterious morphing qualities of the Bt potato—now it's a food, now it's a pesticide—Pollan also discovered the legal loopholes the FDA had to jump through to institute their hands-off policy. "Under FDA law, any novel substance added to a food must—unless it is 'generally [recognized] as safe' ('GRAS', in FDA parlance)—be thoroughly tested. . . . Under the guidelines, new proteins engineered into foods are regarded as additives (unless they're pesticides), but as Maryanski explained, 'The determination whether a new protein is GRAS can be made by the company.' Companies with a new biotech food decide for themselves whether they need to consult with the FDA by following a series of 'decision trees' that pose yes or no questions like this one: 'Does . . . the introduced protein raise any safety concern?'"*

Pollan did run into at least one person who was unaware of the role that the biotech companies had in determining if their own products were safe. The man said his company "should not have to vouchsafe the safety of biotech food. Our interest is in selling as much of it as possible. Assuring its safety is the FDA's job."[36] The man was Phil Angell, Director of Corporate Communications for Monsanto.

Just as the FDA regulates GM foods with lower standards than other food additives, the EPA regulates them with lower standards than chemicals. According to Suzanne Wuerthele, an EPA toxicologist, "For chemicals, we have formal risk assessment guidelines; science policies; conferences where scientific issues are debated." GM foods don't enjoy these safeguards. "We don't even have an understanding of the full range of hazards," says Wuerthele.

She explains, "In the US, each risk assessment for [GM] organisms is done on an ad hoc basis by different scientists in different departments of different agencies. Some of these agencies have conflicting missions— to promote and to regulate; or to consider 'benefits' as well as risks. There is rarely any formal peer review. When peer review panels are put together, they are not necessarily unbiased. They can be filled with [GM] proponents or confined to questions which avoid the important issues, so that a predetermined decision can be justified. This technology is being promoted, in the face of concerns by respectable scientists and in the face of

---

* Many people believe that the FDA policy defines GM foods as "substantially equivalent" to their natural counterparts. While this is true in other countries, it is not the case in the USA. The term exposed the agency's policy to challenges, so they have stopped using it in connection with GM foods.

data to the contrary, by the very agencies which are supposed to be protecting human health and the environment. The bottom line in my view is
that we are confronted with the most powerful technology the world has
ever known, and it is being rapidly deployed with almost no thought
whatsoever to its consequences."[38]

Complete and Accurate Data?

When the FDA first introduced its policy on GM foods, they created a
method by which the biotech companies could voluntarily consult with
the agency. All companies chose to participate, as it was quite superficial.
The *New York Times* described it as a way the companies could "talk to
regulators about the safety of their new genetically engineered products at
least 120 days before they are sold."[39]

In response to public criticism about the regulatory policy on GM foods
and demands by consumer and environmental groups for mandatory
labelling of GM foods, in May 1999 the Clinton administration announced
a set of changes that were meant to bolster consumer confidence. Chief
among these was to make these same consultations mandatory.

Ohio congressman Dennis Kucinich described the meaningless
changes as a way to deflect legitimate concerns about the technology.
"This proposal is full of genetically engineered baloney," said Kucinich,
who described the proposed mandatory consultations as "not worth the
paper they are written on".[40]

Even "FDA officials acknowledged the new rule will mean few, if any,
changes for biofood developers," reported Reuters. "The companies have
considerable freedom to decide what research information and data to share
with the agency. The FDA's mandatory consultations will not affect that."[41]

On January 18, 2001, more than a year and a half after Clinton had
asked the agency to make consultations mandatory, the agency responded
with their proposal. But the agency's version required mandatory "notifi-
cation", not consultation. In other words, companies did not have to talk
with the FDA at all. They could just send in a letter, known as a pre-market biotechnology notice, which describes the food, its method of development, whether it used antibiotic resistant marker genes, information about
substances introduced into the foods (including allergenicity issues), and
some information comparing it to a conventional food.

After another two and a half years of not implementing even this watered down proposal, on June 17, 2003, FDA Deputy Commissioner Lester Crawford told the House Agriculture Committee's subcommittee on research that the agency had decided to abandon the requirement altogether. According to Greg Jaffe, biotechnology director for the Center for Science in the Public Interest, "Under the current system, they [biotech companies] could market something without us even knowing it." Jaffe said, "That is not the best way to ensure the safety or instill consumer confidence in these crops."[42]

By putting companies in charge of determining if their products are safe, apparently the government trusts the private sector to conduct the proper tests and to accurately report any safety issues. A look at the record, however, demonstrates plenty of evidence to the contrary.

In January 1992, months before the FDA policy of self-policing became official, the GAO "claimed that the FDA might be approving drugs for food-producing animals on the basis of 'invalid, inaccurate or fraudulent data' supplied by private laboratories." The GAO said the FDA's "inadequate procedures" could mean that it "may be unable to fulfil its mission to protect the health and safety of animals and people."

This wasn't the first time the FDA had been accused of relying on filtered or flawed industry reports. In 1975, Ralph Moss reported that three pharmaceutical companies had "withheld pertinent information or simply fed the agency false data. . . . Even though either strong administrative sanctions or criminal prosecution might have ensued had FDA so wished, FDA Commissioner Alexander Schmidt told Senate investigators that 'the cases somehow went into some bottomless pit that we have not been able to identify.'"[31]

Also at that time, after studies revealed that the drugs Aldactone and Flagyl "were correlated with cancer in test animals. . . . Ralph Moss reported: 'Further investigation revealed that Searle had known about the tumour-producing potential of these items but had simply given the FDA fraudulent data.'" Searle was the tenth largest pharmaceutical company at the time. It later became a wholly owned subsidiary of Monsanto.

A report by the *Washington Post* revealed that in 1975, while investigating the safety of PCBs, Monsanto's "company study found that PCBs caused tumours in rats. They ordered its conclusion changed from 'slightly tumourigenic' to 'does not appear to be carcinogenic.'"[43]

And in 1990, EPA scientist Cate Jenkins discovered several instances of apparent fraud by Monsanto and urged the agency to do a criminal

investigation. She wrote, "Monsanto has in fact submitted false information to the EPA which directly resulted in weakened regulations." Jenkins
cited internal Monsanto documents that reveal, among other things, that
they doctored samples of herbicides that were submitted to the USDA, hid
evidence, substituted false information "and excluded several hundred of
its sickest former employees from its comparative health studies."[44]
Jenkins said the study by Monsanto "apparently has not been shown to be
a fraud."[45]

Methods of Influence

How does the biotech industry do it? How do they continue to virtually
dictate policy to the US regulatory agencies in spite of such serious and
blatant past transgressions?

Certainly considerable campaign contributions have an influence. The
four leaders of the biotech industry—Monsanto, Dow, DuPont and
Novartis (now Syngenta)—gave more than $3.5 million in PAC, soft-
money, and large individual contributions between 1995 and 2000—
three-quarters of it to Republicans.[46]

In 1994, 181 congressmen co-sponsored a bill that would require
labelling of GM foods. But the twelve-member Dairy Livestock and Poultry
Committee stalled the bill until the end of the 1994 session—effectively
killing it. In testimony before an FDA panel, Robert Cohen said, "I investigated these twelve men and found that collectively they took $711,000 in
PAC [political action committees] money from companies with dairy interests, and four of the members of the committee took money directly from
Monsanto."[47]

Monsanto's Shapiro was among the largest contributors of soft money
to the Clinton re-election campaign in 1996.[48] Shapiro, in turn, became a
member of the President's Advisory Committee for Trade Policy and
Negotiations and served a term as a member of the White House Domestic
Policy Review.[44] And Clinton even praised Monsanto by name in a State of
the Union address.[48]

Lobbying is another way the biotech industry exerts influence.
According to the Center for Responsive Politics, between 1998 and 2002,
the industry spent $143 million on lobbying. This includes the
Biotechnology Industry Organization (BIO), which lobbies and advertises
on behalf of the whole industry. According to a June 2002 Associated
Press report, BIO "has a total budget of $30 million, employs 70 and rep-

resents 1,000 companies." It has made a diverse list of enemies, including the National Right to Life Committee, which accuses BIO of "wielding undue influence on legislation".[49]

"They're everywhere," says Joe Mendelson, legal director of the Center for Food Safety. "The biotech industry is a political force. It's increasing in clout." The organization even runs pro-biotech TV commercials "in Washington, D.C., clearly aimed at legislators considering the issue".[49] In addition, the industry has committed a quarter of a billion dollars over five years to convince the public that GM foods are the right choice.

Perhaps even more important than donations, lobbying and advertising, is the role personal connections play in winning political support. According to the *New York Times*, Monsanto maintains "close ties to policy makers—particularly to trade negotiators". For example, Mickey Kantor, former secretary of the United States Department of Commerce, was a personal friend of Monsanto's CEO Shapiro. Naturally, when Kantor became the trade representative of the United States under Clinton, a strong, sometimes bullying pro-biotech strategy pervaded the US trade stance with the rest of the world. "Confrontation in trade negotiations became the order of the day," reported the *New York Times*. "Senior administration officials publicly disparaged the concerns of European consumers as the products of conservative minds unfamiliar with the science."[1]

(This mindset continues. In March 2003, Speaker of the House Hastert lashed out at the European Union's "protectionist, discriminatory trade policies" on GMOs, which the Speaker called "non-tariff barriers that are based on fear and conjecture—not science".[50])

After leaving government office, Mickey Kantor became a member of Monsanto's board of directors. Another official-turned-board-member was William Ruckelshaus, the former chief administrator of the EPA. The *Globe and Mail* describes Monsanto as "a virtual retirement home for members of the Clinton administration".[48]

Another former EPA employee, Linda Fisher, became vice president of government and public affairs for Monsanto before returning to the EPA to become their second in command. Lidia Watrud, former biotechnology researcher at Monsanto, joined the EPA's Environmental Effects Laboratory.

At the FDA, two former Monsanto employees along with Michael

134 SEEDS OF DECEPTION

Taylor approved Monsanto's genetically engineered bovine growth hormone—which no other industrialized nation has yet deemed safe for their cows or their milk-drinking population. Taylor, by the way, is a cousin of Al Gore's wife Tipper Gore. "The Food and Drug Administration," says Betty Martini of the consumer group Mission Possible, "is so closely linked to the biotech industry now that it could be described as their Washington branch office."[48]

To navigate Washington's complex bureaucracy, Monsanto looks to their director of international government affairs, Marcia Hale. She had been assistant to the president of the United States and director for intergovernmental affairs. Similarly, Monsanto's director of global communication, Josh King, was formerly the director of production for White House events.[51]

Some other strategic job swaps between the biotech industry and government: Genentech's David W. Beier became Vice President Al Gore's chief domestic policy advisor. Clayton K. Yeutter, former Secretary of Agriculture and former US trade representative, became a member of Mycogen's board of directors. L. Val Giddings, Vice President of BIO, was biotechnology regulator and (biosafety) negotiator at the USDA. And Terry Medley, Dupont's director of regulatory and external affairs, held senior positions at the USDA and FDA.

Leading figures in the George W. Bush administration also have significant ties to the biotech sector. Agriculture Secretary Ann Veneman was an attorney with a firm that represented biotech corporations. She was also on the board of Calgene, Inc., now a subsidiary of Monsanto.[52] Defense Secretary Donald Rumsfeld was the president of another Monsanto subsidiary, Searle—makers of the genetically engineered sweetener aspartame. Mitch Daniels, director of the office of management and budget, was vice president at Eli Lilly Pharmaceutical company, partners with Monsanto on the genetically engineered bovine growth hormone. Tommy Thompson, secretary of health, received $50,000 from biotech firms during his earlier Wisconsin gubernatorial election. Thompson used state funds for a $317 million dollar biotech zone in the state.[53] John Ashcroft, attorney general, was the largest recipient of campaign funds from Monsanto in the 2000 election, while Supreme Court Judge Clarence Thomas once worked as Monsanto's lawyer.

Journalist Bill Lambrecht described an example of how Washington's biotech connections came into play during a carefully orchestrated a 1998

St. Patrick's Day reception for the Irish prime minister, Bertie Ahern. His vote was needed to carry the EU's acceptance of Monsanto's GM maize. When Ahern had lunch with National Security Advisor Council Director Sandy Berger, the topic that Berger chose to focus on was the need to get that maize vote. Again, when Ahern met Senator Bond from Missouri and several members of congress, the issue was GM maize. According to Toby Moffet a former congressman turned Monsanto man, "Everywhere he went, before people said 'Happy St. Patrick's Day,' they asked him, 'What about that corn vote?'" The amazed Moffet said, "I'm fifty-four years old, and I've been in a lot of coalitions in my life, but this is one of the most breathtaking I've seen."

The next day, Ireland cast its vote in favour of Monsanto's GM maize, the first time Ireland acted in favour of a GMO release. When revelations of the events in Washington were made public by Lambrecht in the *St. Louis Post Dispatch*, the Irish group Genetic Concern charged in a press release, "US multinationals have more influence than the Irish electorate."[54]

Moderate Dissent among the Ranks

Former Secretary of Agriculture Dan Glickman had been one of the Clinton administration's staunchest defenders of biotech, touring Europe with industry representatives to promote GM foods. In an interview just before stepping down from office, he said:

> "What I saw generically on the pro-biotech side was the attitude that the technology was good, and that it was almost immoral to say that it wasn't good, because it was going to solve the problems of the human race and feed the hungry and clothe the naked. . . . And there was a lot of money that had been invested in this, and if you're against it, you're Luddites, you're stupid. That, frankly, was the side our government was on. Without thinking, we had basically taken this issue as a trade issue and they, whoever 'they' were, wanted to keep our product out of their market. And they were foolish, or stupid, and didn't have an effective regulatory system. There was rhetoric like that even here in this department. You felt like you were almost an alien, disloyal, by trying to present an open-minded view on some of the issues being raised. So I pretty much spouted the rhetoric that everybody else around here spouted; it was written into my speeches."[55]

In 1999, however, Glickman broke ranks with the pro-biotech hardliners of the Clinton administration, albeit cautiously. In a speech at Purdue University, he said the United States "can't force-feed consumers" around the world. And in a speech at the Press Club in Washington, "Glickman advised biotechnology companies to consider labelling genetically modified food to help prevent consumer fears from spreading to the United States."[55]

According to the *St. Louis Post Dispatch*, "That was not what the heavily invested industries—or the White House, for that matter—had expected him to say. He purposely had not submitted his speech for approval beforehand, he recalled, because he knew it would be returned to him 'sterile'. Afterward, he felt the heat." Glickman said, "There were some people in this government who were very upset with me. Very upset."

When Glickman met the president's wife, Hillary Rodham Clinton, at a White House dinner a few days after his speech, Glickman later reported: "She said, 'I saw the story about your speech in the *New York Times*.' I said to her, 'There were some people in the White House that didn't like it.' She said, 'I liked it.' So I knew I wasn't going to be fired."

Glickman's concerns about GM foods run deeper than just labelling. He wants "a thorough review of how GMOs are regulated by our government". He says, "I think it does need further clarity."[56]

Where Has the Government's Push Got *Them?*

Dan Quayle's Council on Competitiveness de-regulated GM food in order to strengthen the economy and make American products more competitive overseas. In the decade since, here is what has happened.

Major retailers and food manufacturers around the world responded to consumer pressure by vowing to remove GM ingredients from their brands. In Europe, nearly the entire food manufacturing and retail industry has banned GM ingredients, and the majority of the world's population are covered by restrictions on the sale and use of GM crops.[57]

Because of the difficulty of segregating GM crops from non-GM crops, many overseas buyers have simply rejected all maize, soya, oilseed rape and cotton from the US and Canada. Since these four GM crops and their derivatives are found in most processed foods in the US, American-made packaged foods are also off-limits in many markets.

US maize exports to Europe have been virtually eliminated, down by 99.4 percent. Likewise, Canada's annual oilseed rape sales to Europe vanished as has their honey—tainted by GM pollen.[58] US soya, which enjoyed

57 percent of the world market, dropped by about a fifth to 46 percent.[59] Soya is principally used for animal feed. US soya sales have been supported by the fact that, until recently, few importers expressed concern about feeding animals GM feed. But overseas retailers are now promising to sell meat that was raised on non-GM sources. The USDA announced in May 2001 that European demand for non-GM feed jumped from near zero to 20 to 25 percent within twelve months.[58]

The lost markets for US crops caused near record low prices. The American Corn Growers Association (ACGA) calculated the resultant drop in maize prices at 13 to 20 percent.[57] According to Charles Benbrook, former executive director of the National Academy of Sciences' Board on Agriculture, growers have only been kept afloat by the huge jump in subsidies.[60] Benbrook estimates government payments to farmers are up by $3 to 5 billion annually due to the economic damage of GM crops alone.[57]

Although GM tomatoes and potatoes failed to take hold and have been taken off the market, and GM sugar beet, flax and rice, while approved, were never commercialized, Monsanto is pushing hard to introduce GM wheat. Two-thirds of US and Canadian wheat buyers, however, don't want GM wheat and may go elsewhere if it is introduced. According to the UK Soil Association's report, *Seeds of Doubt*, "The US and Canadian National Farmers Unions, American Corn Growers Association, Canadian Wheat Board, organic farming groups and more than 200 other groups are lobbying for a ban or moratorium on the introduction . . . of GM wheat."[57]

Even in the US where there has been far less news coverage of the GMO issue, more and more food manufacturers are committing to remove GM ingredients from their products. "First major health food retail chains such as Whole Foods and Wild Oats rejected GMOs. Now mainstream American retailer Trader Joe's has followed suit as a result of market research: 'The majority of our customers would prefer to have products made without genetically engineered ingredients.' Other, even larger US-based food companies, including Frito-Lay, Gerber, Heinz, Seagram and Hain, have also decided not to use GMOs in their products. A study by Rutgers University Food Policy Institute in November 2001 also revealed that the vast majority of the US population want GM food to be labelled."[57]

Stuck with products no one wants, the US has tried to give GM grain away as food aid to developing nations. But consumer groups and governments alike regularly reject the food, which they say has not been proven safe.

So instead of creating a solution to the trade deficit, GM crops have been a disaster for US trade. "In total, with the lower profitability of GM crops, the loss of foreign trade, the lower market prices, the costs of the StarLink maize recall and other incidents, the farm subsidy rise and the lost . . . organic market opportunities, GM crops could have cost the US economy some $12 billion net from 1999 to 2001."[61]

When Robert Shapiro shifted Monsanto's strategy to the fast track, he predicted rapid, global acceptance of GM crops. Although the top biotech companies own 23 percent of the commercial seed market and total GM acreage far exceeds the size of the UK, many observers agree that Monsanto's push of genetically engineered foods has been a failure. The company's aggressive strategy has been credited, in part, for the eruption of global opposition to GM foods.

Shapiro confessed to a Greenpeace gathering in October 1999 that Monsanto "irritated and antagonized people".[62] Will Carpenter, who headed Monsanto's biotechnology strategy group until 1991, describes it more eloquently. "When you put together arrogance and incompetence, you've got an unbeatable combination. You can get blown up in any direction. And they were."

The US government, however, continues to echo the gung-ho attitude that Glickman describes. They blame anti-GMO sentiment largely on baseless, irrational fears. According to the book *Trust Us We're Experts*, "Government and industry insiders rationalize the gulf that separates them from popular opinion by dismissing citizen concerns with the usual rhetoric about the public's ignorance. Terms such as 'Luddite' and 'looney' abound as the biotechnicians compete among themselves to see who can express the most contempt for the intelligence of the great unwashed masses."[63]

US Deputy Secretary of Commerce David Aaron told European representatives in 1999, "Not a rash, not a sneeze, not a cough, not a watery eye has been developed from [GM foods], and that's because we have been extremely careful in our process of approving them."

He said that the FDA found no scientific proof that GM foods were harmful. He said the reason that Americans were not against GM food is because they trust the FDA. The problem, according to Aaron, was not with the foods. It was the fact that Europe had no American-style FDA. Aaron said, "We would like the governments . . . to develop a transparent, systematic approval process that is based on science."[64]

That would be nice.

Wisdom of the Rats

The *Washington Post* reported that rodents, usually happy to munch on tomatoes, turned their noses up at the genetically modified FlavrSavr tomato that scientists were so anxious to test on them. Calgene CEO Roger Salquist said of his tomato, "I gotta tell you, you can be Chef Boyardee and . . . [they] are still not going to like them."[1]

Rats were eventually force fed the tomato through gastric tubes and stomach washes. Several developed stomach lesions; seven of forty died within two weeks. The tomato was approved.

Chapter 6

Rolling the Dice
with Allergies

When her one-year-old daughter developed an allergy to milk in February 1998, a leading British surgeon did what many other mothers do: she switched to soya milk. When the girl immediately developed large cold sores, the child was tested and found not to be allergic to soya. The mother figured it must be something else and continued feeding her soya milk. Over the next year the sores got worse and did not respond to treatment. "I became aware that she was not getting better," said the mother. "There seemed to be three large, weeping sores on her face at any one time."

From a geneticist friend, she learned about the potential risks of GM soya and tried reducing the daughter's soya milk by one-fourth. "The sores cleared up overnight," she recalled.

She told the *Sunday Telegraph*, "I want the government to look into this because I saw the change in my daughter—as soon as she was taken off the GM milk, her health dramatically improved. I and my [general practitioner] have not found any other reasons why she became ill. My family previously ate GM products without worrying—but now we do not."[1]

Could the child have reacted to the GM soya but not natural soy? It's possible, but the limited details raise more questions than they answer. Did the allergy test use natural soya instead of GM soya, thereby missing her reaction to the GM variety? Was the reaction not an allergy but rather a food "intolerance" or "sensitivity" to GM soy? The mother's geneticist friend even suggested that the cold sores were related to a virus that was being activated by the GM soy.

If GM soybeans were responsible for anything out of the ordinary such as increased allergies, then the total number of allergies attributed to soya would probably rise in the general population after GM soya was introduced into the diet. Unfortunately, very few countries maintain detailed statistics on food allergies. In the UK, however, the York Nutritional Laboratory, Europe's leading specialist on food sensitivity, does extensive tests each year to determine how many people have allergies and to what foods.

In March 1999, York Laboratory scientists discovered that soya allergies skyrocketed over the previous year, jumping 50 percent. The increase propelled soya into the top ten list of allergens for the first time in the seventeen years of testing. Soya "moved up four places to ninth and now sits alongside foodstuffs with a long history of causing allergies, such as yeast, sunflower seeds and nuts," reported the *Daily Express*.

Researchers tested 4,500 people for allergic reactions to a wide range of foods. In previous years, soya affected 10 percent of consumers. Now, 15 percent reacted with a range of chronic illnesses, including irritable bowel syndrome, digestion problems, and skin complaints including acne and eczema. (Note: Some of these reactions may fall into the category of food hypersensitivity or food intolerance, not food allergies per se. For the sake of this discussion, we will not distinguish between the categories.) According to John Graham, spokesman for the York laboratory, "People also suffered neurological problems with chronic fatigue syndrome, headaches and lethargy." Scientists confirmed the link with soya by detecting increased levels of antibodies in the blood. Furthermore, the soya tested in the study, like most soya in the UK at the time, was primarily imported from the USA and therefore contained a significant percentage of the genetically modified Roundup Ready variety.

The fact that GM soya had recently entered the food supply was not lost on the researchers who, according to the *Daily Express*, "said their findings provide real evidence that GM food could have a tangible, harmful impact on the human body." Graham said, "We believe this raises serious new questions about the safety of GM foods."

The British Medical Association had already warned that the technology may lead to the emergence of new allergies. With York's research in hand, British scientists now urged their government to impose an immediate ban on GM foods until further testing evaluated their safety. Irish doctors also demanded that GM foods be banned, when increased soya allergies were also reported in that country. Geneticist Michael Antoniou said that the increase in allergic responses "points to the fact that far more work is needed to assess their safety. At the moment no allergy tests are carried out before GM foods are marketed."[2]

Soya and soya derivatives are used in more than 60 percent of processed foods sold in the USA. GM soya is mixed with natural soya and foods are not labelled as such. Avoiding GM soya, therefore, is a difficult task.

There are many potential reasons why GM soya could be allergenic. Increasing the amount of a naturally occurring plant allergen is one way that genetic modification might promote allergies. Trypsin inhibiter, a substance found in natural soya, has been identified as a major allergen. According to a published study, the amount of trypsin inhibitor in one variety of GM soybeans is about 27 percent higher than in natural soybeans.[3] It is also possible that GM food possesses new allergens, never before found in natural food.

Transferring Allergens

Researchers at Pioneer Hi-Bred, a leading US seed company now owned by Dupont, wanted to genetically engineer a soybean that would be a more "complete protein" for animal feed. They needed to "borrow" an amino acid somewhere. Their final candidate: a gene from the Brazil nut. When they inserted it, their soybean did acquire the desired trait—extra nutrition for the diets of cows and pigs alike.

Before sending it to market, they decided to test the bean for possible allergenic effects. They knew that some people are allergic to Brazil nuts, in rare cases fatally so. And although it was created with animals in mind, the bean would also end up in the human diet.

They contacted University of Nebraska scientist Steve Taylor who, according to the *Washington Post*, "practically yawned when [Pioneer Hi-Bred] asked him in 1995 to study a new soybean they had invented. 'I didn't think we'd find anything interesting,' Taylor recalled."

He reasoned that Pioneer had taken only one protein out of the thousands of proteins found in a Brazil nut. The odds of that one being the source of the nut's allergenicity were incredibly low. Taylor was therefore amazed when three separate tests demonstrated that the soya did in fact cause reactions in people allergic to Brazil nuts. "In trying to build a better soybean," reported the *Washington Post*, "the company had made a potentially deadly one."[4] According to the article, this study was "one of the very few studies ever to look directly for any harm from an engineered food or crop." When it was eventually published in the *New England Journal of Medicine*,[5] the biotech industry and the world were put on notice about a serious potential danger of genetic engineering.

To guard against this danger, the FDA's 1992 policy lists examples of foods with known allergens and indicates that if a GM food uses genes from any of these, the manufacturer should consult with the agency. The allergens in their list—milk, eggs, fish, shellfish, nuts, wheat and legumes—account for about 90 percent of American food allergies. The many foods responsible for the other 10 percent of reactions are not included. FDA toxicologist Louis Pribyl was apparently unhappy about this omission. In a March 1992 critique of an early draft of the policy, he wrote, "there are very few allergens that have been identified at the protein or gene level." Biotech companies could therefore never be sure if their GM crops were free of transferred allergens. He said, "Companies are going to have to consult FDA" not just on genes taken from the common allergenic foods, but "every other plant which produces allergic reactions".⁶ His recommendation was not adopted.

The policy specifically states, "Producers of such foods [with known allergenicity] should discuss allergenicity testing protocol requirements with the agency."⁷ While the wording suggests that tests are required, James Maryanski of the FDA explains that the agency actually offers suggestions; any testing is voluntary.⁸ The policy continues, "labelling of foods newly containing a known or suspect allergen may be needed to inform consumers of such potential."⁷ Once again, this is only a suggestion. Critics argue that without labelling, not only would people be susceptible to allergic reactions, they might never know what caused their reaction and how to avoid it in the future. "This lack of predictability is worrying for people with food allergies," says the UK magazine *GM-Free*. "These people can only live their lives on the basis that they know which foods to avoid."⁹

More worrisome is that current GM foods get their genes from bacteria, viruses and other organisms. No one knows if humans are allergic to their proteins—they were never before part of the human food supply. According to the FDA's 1992 policy, "At this time, FDA is unaware of any practical method to predict or assess the potential for new proteins in food to induce allergenicity and requests comments on this issue."⁷ According to a 1999 *Washington Post* article—written seven years later—there is still "no widely accepted way to predict a new food's potential to cause an allergy. The FDA is now five years behind in its promise to develop guidelines for doing so. With no formal guidelines in place, it's largely up to the industry to decide whether and how to test for the allergy potential of new food not already on the FDA's 'must test' list."⁴

A 1996 editorial in the *New England Journal of Medicine* said, "Because FDA requirements do not apply to foods that are rarely allergenic or to donor organisms of unknown allergenicity, the policy would appear to favour industry over consumer protection."[10]

The FDA does recommend that producers evaluate potential allergens by comparison of the protein's amino acid sequence to known allergens, the resistance of the protein to break down by digestion and heat, and evaluation of molecular size. The EPA, which regulates pesticidal Bt crops, makes similar recommendations. Most scientists agree, however, that these are unreliable methods and cannot fully safeguard the public.

"None of these criteria are exact," said Hansen, "as the state of science in the field of allergenicity is still in its infant stages."[11] Arpad Pusztai describes the FDA's allergy test methods as indirect and rather scientifically unsound. The FDA's own scientist Carl Johnson writes, "Are we asking the crop developer to prove that food from his crop is non-allergenic? This seems like an impossible task."[12]

New foods are very difficult to test for allergenicity. People aren't usually allergic to a food until they have eaten it several times. According to FDA's Pribyl, "the only definitive test for allergies is human consumption by affected peoples, which can have ethical considerations."[6] Pusztai concurs, saying, "It is at present impossible to definitely establish whether a new GM crop is allergenic or not before its release into the human/animal food/feed chain."[13] He said, "I think that is the Achilles heel of these GM foods. What they do now in testing is rubbish."[14]

This has led some scientists to call for "post-market surveillance" of new GM foods for allergic reactions, in much the same way newly introduced drugs are monitored for side effects.[15] The UK Royal Society recommends such surveillance in particular for "high-risk groups such as infants".[16]

In January 2001, the Food and Agriculture Organization of the United Nations (FAO) and the World Health Organization (WHO) convened a joint expert consultation and created a set of recommended guidelines to evaluate the allergenicity of GM foods. While they acknowledge that it is impossible to predict allergies with certainty, they created a series of questions in a decision tree format to better determine if a GM food will cause an allergic reaction.

While both the FDA and EPA acknowledge that better allergy testing is needed, they have not embraced the FAO/WHO guidelines, which are more

stringent and comprehensive than the agencies want. In fact, currently registered Bt crops would likely fail the FAO/WHO testing protocol.[17] This might
help explain why US regulators are attempting to promote less strict criteria,
that will be less of a burden on the industry.

According to the *New England Journal of Medicine*, about one quarter of Americans "believe that they or their children are allergic to specific
foods".[10] Blood test results, however, indicate that the confirmed number
is more like 2 to 2.5 percent of adults and up to 8 percent of children—
about eight million Americans. For unknown reasons, allergies are on the
rise. The StarLink incident of 2000 demonstrates how GM foods might be
contributing to the increasing number of allergies, and how unprepared
the government is to monitor, detect, or deal with allergic outbreaks.

StarLink Shock

At a business lunch with co-workers in September 2000, thirty-five-year-
old Grace Booth dined on three chicken enchiladas, which she later
recalled were very good. Within about fifteen minutes, however, something went wrong. She felt hot, itchy. Her lips swelled; she lost her voice
and developed severe diarrhoea.

"I felt my chest getting tight, it was hard to breathe," recalled Booth.

"She didn't know, but she was going into shock," reported CBS news.

"I thought, oh my God, what is happening to me? I felt like I was
going to die."[18] Her co-workers called an ambulance.

In the emergency room of a nearby hospital in Oakland, California,
Booth was injected with anti-allergy medicine, given Benadryl, and put on
an IV. It worked. The effects of anaphylactic shock subsided and five
hours later Booth safely left the hospital.

Across the country, Keith Finger, a Florida optometrist, enjoyed a dinner of tortillas, beans and rice. Fifteen minutes later he got a terrible stom-
achache and diarrhoea. Soon he was itching all over. His tongue started to
swell and he had trouble breathing—again the symptoms of anaphylactic
shock. Finger injected himself with anti-allergy medicine and swallowed
some Benadryl; the symptoms subsided. He is confident, however, that
without the medicine he would have died.

Neither Booth nor Finger knew what had caused their allergic reactions, but within a few days both heard the news. A genetically modified
maize product called StarLink, which contained a potential allergen and
was not approved for human consumption, was discovered in tacos, tor-

tillas and other maize products. More than 300 items were eventually recalled from the grocery store shelves in what was to become the world's biggest GM food debacle.

Booth contacted the Food and Drug Administration. There was maize in her tortillas and she had tested negative for all other food allergies. Booth thought StarLink might be the cause. Finger too confirmed that there was maize in his tortillas and filed a report with the FDA.

Hundreds of others also contacted the FDA, concerned that they too had allergic reactions to StarLink; more than fifty people eventually filed reports with the agency. Symptoms "varied from just abdominal pain and diarrhoea [and] skin rashes to . . . a very small group having very severe life-threatening reactions," said Marc Rothenberg, chief allergist at Cincinnati Children's Hospital and adviser to the government in the StarLink investigation. Twenty-eight people's reaction fit the profile of an anaphylactic response.[19]

StarLink was not supposed to be eaten by humans. It is a brand of maize that creates a modified form of a pesticide produced by the soil bacteria *Bacillus thuringiensis* (Bt). But StarLink creates a version of the toxin that is unlike the toxin in other Bt maize varieties. The StarLink version is called Cry9C. "StarLink is suspected of causing allergies because Cry9C has a heightened ability to resist heat and gastric juices—giving more time for the body to overreact," reported the *Washington Post*. This property was created intentionally, as a means to enhance the maize's ability to kill pests. In addition to having longevity in the digestive tract, StarLink protein's molecular weight is "consistent with something that can trigger an allergic reaction".[20] The Environmental Protection Agency (EPA), which oversees GM crops that create their own pesticide, therefore, did not approve the maize for human consumption.

(It is interesting to note that the FDA did not express concern about StarLink. In a May 29, 1998 letter, the FDA wrote to AgrEvo [the company that developed StarLink, later purchased by Aventis], "Based on the safety and nutritional assessment you have conducted, it is our understanding that AgrEvo has concluded that maize grain and forage derived from the new variety are not materially different in composition, safety, or other relevant parameters from maize grain or forage currently on the market, and that do they do not raise issues that would require pre-market review or approval by FDA."[21] Note that the FDA here relies entirely on the company's own safety assessment, as it does for all GM crops.)

The EPA, however, did allow StarLink to be fed to pigs, cows and other livestock. The EPA also required that the manufacturer let farmers know that the maize must be segregated. Farmers were supposed to sign statements that any StarLink they grew, plus any maize grown within 660 feet of it, was only to be used for animal feed or industrial (fuel) purposes, but not put into the human food chain.

In spite of these requirements, the word about the maize's special handling instructions didn't circulate much. Farmers didn't know; grain elevators didn't know. In fact, some StarLink seed tags explicitly stated that the maize was suitable for "forage or grain for food, feed or grain processing".[22] Therefore, although StarLink was planted on less than 1 percent of US cornfields—312,000 acres—it was readily mixed in grain silos across the United States, contaminating 22 percent of the grain tested by the USDA. Some proportion of StarLink was eventually found in tacos, corn chips, corn meal, and all things corn. Over 10 million individual food items were subject to recall, but not before tens of millions of people had eaten StarLink in their diet.[23]

The StarLink problem was a huge setback for the biotech industry. The US public began questioning the safety of GM foods for the first time. The government came under fire for approving maize for animals and not for humans, knowing that the grain processing system in the US is not equipped for such segregation. US maize exports and prices plummeted, as major trading partners like Japan and Korea looked elsewhere for maize that was free of StarLink contamination. The event also threatened to drive a wedge between the biotech industry and the US food industry, which had to deal with product recalls, brand name damage and consumer fears.

The Elusive Allergy Test

With consumers concerned about their health and US maize exports declining, the FDA was under intense pressure to determine whether StarLink was, in fact, an allergen. At the same time, the agency was "up against the reality that there is no sure-fire way of testing a new protein like Cry9C for its potential to cause allergies in people," reported the *Washington Post*. "We all wish there was a test where you plug in a protein and out pops a 'yes' or 'no' answer," said Sue MacIntosh, a protein chemist with AgrEvo. "But there is no such test . . . short of giving it to a lot of people and seeing what happens."[4]

After months of waiting, the FDA and the Center for Disease Control (CDC) came up with a plan for an allergy test. Karl Klontz, a medical officer with the FDA's Center for Food Safety and Applied Nutrition, said, "This is the first time a test like this has been developed, and nobody is claiming that it is a gold standard." The *Washington Post* reported, "It has not been fully checked and double-checked and researchers warn the test will not give a definitive answer."[4]

The FDA's test involved looking for antibodies in blood samples from seventeen people who were suspected of being allergic to StarLink—they had reported serious allergic reactions after eating maize products and were not normally allergic to maize. The presence of antibodies would indicate that some reaction to Cry9C had taken place. Based on the results, on June 11, 2001, nine months after Booth ate her enchiladas, the FDA announced the test results: StarLink was not the cause of allergies. The biotech industry was quick to disperse the news, claiming as always that GM food was safe to eat. Val Giddings of the Biotechnology Industry Organization said that the results meant that the case was "slam-dunk closed".[24] But as the details of the FDA test emerged, scientists became critical of its design and suspicious of its conclusions.

Just five weeks after the FDA/CDC's declaration of safety, advisers to the EPA—including some of the nation's leading food allergists—released a thorough critique of the FDA's allergy test and other aspects of the StarLink investigation. Their conclusion? "The test, as conducted, does not eliminate StarLink Cry9C as a potential cause of allergic symptoms."[25] They said the research had many shortcomings. For example, the test lacked adequate controls, was not sensitive enough, and failed to follow standard protocols that helped prevent false interpretations.

Perhaps the gravest error was that the FDA asked Aventis, the makers of StarLink, to provide the Cry9C. If the FDA was under significant pressure to create a StarLink test, Aventis was under far more pressure to pass that test.

The StarLink mishap was already mounting a hefty price tag for this Swiss-based corporation. Although they had agreed to buy back farmers' StarLink inventory for a price tag of at least $90 million, they were facing at least nine class actions from individuals and companies attempting to recover millions in losses. These included:

- Farmers who lost money because their maize was mixed with StarLink.
- Farmers who suffered from lost markets and falling prices. (US maize exports dropped by fifty million bushels or more and maize prices hit their lowest levels in about fifteen years.)
- Consumers who claimed allergy related problems.
- Companies who recalled more than 300 products.
- Taco Bell franchises and other Mexican food companies who claimed reduced business due to maize fears.

In addition to lawsuits, Aventis received "hundreds of angry phone calls from farmers, grain elevator managers and food processors".[26] Eighty-seven of its employees re-routed 28,135 trucks, 15,005 rail cars and 285 barges to limit the chances that StarLink would mix with maize destined for human consumption. Aventis' eventual price tag for the StarLink contamination is estimated at $1 billion.

In an attempt to limit their damages, Aventis petitioned the EPA to declare the remaining StarLink in the food supply legal. They claimed StarLink was safe, and that there was so little of it left in the food supply that even if it was allergenic the amount would be too low to create an effect. They also admitted that due to cross-pollination and other factors, StarLink would remain in the food supply forever.

With Aventis under such intense pressure, it seemed an odd choice to ask them to hand over the protein that would be used in the allergy test to clear its name. According to the Friends of the Earth, "There is no evidence that the FDA made any attempt to independently verify the composition or purity of the Aventis-supplied" protein or antibodies. " 'Conflict of interest' is apparently not a concept the agency is familiar with."[23] The EPA's Scientific Advisory Panel agreed. They wrote: "Aventis appears to have furnished all samples for the current evaluation. Is this appropriate? The Panel favours establishment of a procedure to independently validate reagents and materials."[25]

Aventis gave the FDA a sample of Cry9C protein, but it wasn't taken from StarLink. Claiming that they couldn't isolate enough of the protein from the maize, they offered a synthesized protein substitute derived from *E. coli* bacteria. Interestingly enough, even the FDA admits that this substitution could invalidate the test's results.[27]

Substituting protein derived from *E. coli* is not unprecedented. Since it provides a less expensive way to produce sufficient quantities of the protein, biotech companies rely on *E. coli*-created proteins for most of their tests. The National Academy of Sciences, however, recommends that GMO tests be conducted with protein produced in the plant, not in the bacteria. They said any bacterial substitute must meet "scientifically justifiable criteria for establishing biochemical and functional equivalency" to the plant-produced protein. Similarly, an expert committee of the European Commission said that such a substitution "can be accepted only if the chemical identity . . . of the two proteins has been demonstrated."[23] Scientific advisers to the EPA have recommended such criteria,[28] but neither the EPA nor FDA has bothered to adopt them.

These guidelines are based, in part, on the principles discussed in Chapter Two. There it was pointed out that same protein is not necessarily identical in different species. They can have different added molecules (hitchhikers), for example, or be folded differently. These were precisely the concerns expressed by the EPA's Scientific Advisory Panel (SAP). Their report said, "There is no assurance that bacteria-derived Cry9C is properly folded."[25]

More importantly, Cry9C created from StarLink maize has an added sugar chain, a hitchhiker, which "is well known to enhance allergenicity of a protein." The Cry9C made from *E. coli*, however, does not have the sugar chain; that may explain why it did not react with the blood from seventeen who claimed to be allergic to StarLink. The EPA had asked Aventis in 1997, long before the StarLink crisis, to determine the composition of the sugar chain in order to assess its allergenicity. Aventis responded that research was underway, but they never reported the results to the agency.

What Aventis did present to a July 2001 meeting of the EPA's Scientific Advisory Panel was wholly inadequate. Mistakes in the document obscured the results, conclusions were at odds with the study's own data, and Aventis failed to update a five-year-old test with newer more reliable methods. Moreover, it took the company eight months to deliver it to the panel.

One frustrated panel member, Dean Metcalfe, M.D., who heads the National Institutes of Health Laboratory of Allergic Diseases, and is the government's top allergist, made the comment, "It is important, I think, for people listening to this to understand that the questions that we have are not really minor questions. To try to put this in perspective, most of us review for a lot of journals. And if this were presented for publication in the journals that I review for, it would be sent back to the authors with all of these questions. It would be rejected."[29] The poor quality of the data made it difficult for the panel to evaluate the sugar chain hitchhiker.

Masaharu Kawata, a Japanese scientist who had conducted his own critical analysis of the StarLink test, said, "We have found many examples of this kind of data comparison that are incomparable and may look scientific, and is the same disguised tactics used in the application for approval of Roundup Ready Soybean[s] by Monsanto in Japan."[30]

Kawata pointed out an additional flaw of the study that he says was "indisputably a mistake". The researchers used as their control group twenty-one blood samples that had been frozen since before 1996—before its donors could have been exposed to StarLink and developed an antibody reaction. This control blood served as the baseline; in order for StarLink to be considered allergenic, reactions to Cry9C by the blood of the seventeen people who had suffered allergic responses to maize needed to be at least 2.5 times greater than the reactions by the control blood. But when tested, reactions in the previously frozen control blood varied widely and were not reproducible. Moreover, the control blood reacted far more to the Cry9C than did the blood from the allergic group. No one knew why this happened but, according to Kawata, the "CDC, after apparent brain racking, came up with an excuse that the blood serum, [which] had been freeze preserved . . . could be different from that of fresh blood samples." Kawata says that this obviously should have disqualified the controls. But researchers stuck by their quirky frozen blood, and since it reacted more to Cry9C than the test groups, StarLink was off the hook.

In the end, after careful analysis of all the available data, however, the EPA's Scientific Advisory Panel upheld their original assessment that there is a medium likelihood that StarLink was an allergen. The Panel also decided that even the twenty parts per billion tolerance for Cry9C requested by Aventis should not be granted. They said, "based on reasonable scientific certainty, there is no identifiable maximum level of Cry9C protein that can be suggested that would not provoke an allergic response and thus would not be harmful to the public." EPA's Stephen Johnson summed up the situation as follows: "Some of the world's leading experts on allergenicity and food safety told us there was not enough data to conclude with reasonable certainty that there was an acceptable level of [StarLink corn] that people could eat." He said it "would require many months or years of continued scientific evaluation to answer the question of allergenicity."[31]

Unfortunately, the EPA has not followed through on the Panel's recommendations for further research.

The advisory panel also recommended to the EPA that allergy testing should be expanded to include all GM foods. According to the *Washington Post*, the panel also said "that 'every attempt' should be made to further test two people who reported severe reactions and who have offered to undergo skin testing and to eat StarLink products under medical supervision."[4]

Dr. Finger, the Florida optometrist who had nearly died after eating a tortilla, had already offered to eat StarLink maize to see if he would have a second reaction. Although risky, this method offers significant advantages over the FDA's methods. Suppose, for example, that the process of genetic engineering had given rise to some of the other unpredictable effects discussed in Chapter Two. Code scramblers, damaged DNA, gene silencing, genetic instability and haphazard promotion by the CaMV promoter, can all potentially change the expression of the natural proteins in maize, or even introduce a new unexpected protein. Even if the FDA's test had not used Cry9C from bacteria, but rather isolated Cry9C from StarLink, testing the protein and not StarLink maize itself might have missed detection of other possible allergens created in the maize. When contacted by Finger with his proposal, Aventis' lawyer "was initially interested but declined".[32]

Nonetheless, after going public with his offer to be tested, he received some StarLink sent to him anonymously in the mail. "After running a test that showed it was in fact StarLink, he ate some and went to a local hospital several hours later with itchy rashes over his body and fast-rising blood pressure," reported the *Washington Post*. Finger's blood had been used as one of the seventeen that had tested negative in the FDA's test.

Friends of the Earth, the organization that had spearheaded the initial discovery of StarLink contamination in the food supply, wrote an analysis of the way the StarLink investigation was carried out. They point out several errors by the government as well as numerous ways in which Aventis failed to cooperate.

For example, the FDA established a passive monitoring system, contacting and testing only the tiny percentage of affected people who filed formal complaints with them. They didn't investigate the thousands of allergy or health-related consumer calls made to food companies, including some who were rushed to emergency rooms.

The FDA did not actively contact health professionals or allergy groups around the country once the contamination had been made public. Maize is not normally considered a major allergen. Eighty percent of the US population eats some form of maize protein every day. Without adequate education, many Americans might have suffered reactions without knowing the cause or how to prevent future problems.

The FDA should have made efforts to protect children, who are three to four times more prone to allergies than adults. Infants below two years old are at greatest risk—they have the highest incidence of reactions, especially to new allergens encountered in the diet. Children generally eat a higher percentage of maize in their diet, and allergic children in particular often rely on maize protein. Even tiny amounts of allergens can sometimes cause reactions in children. Breast fed infants can be exposed via the mother's diet, and fetuses may possibly be exposed in the womb. Mothers using corn starch as a talc substitute on their children's skin might also inadvertently expose them via inhalation.

Friends of the Earth charged Aventis with improperly testing and reporting the properties of their own product. To evaluate how much Cry9C protein remained in the maize after cooking, Aventis heated its maize four times longer than the standard period. This is reminiscent of the way milk from rbGH-treated cows was over-pasteurized in order to try to destroy the growth hormone. In addition, Aventis used shorter-than-recommended protein extraction times, which can also reduce the amount of protein detected.

Also, the company consistently failed to provide critical information about the allergenicity of the product. Even before the contamination was discovered, an EPA Scientific Advisory Panel had asked Aventis to provide blood from animals fed StarLink and from humans who might have been sensitized by inhaling its pollen. They also asked that Aventis monitor agricultural workers who had the greatest exposure to StarLink and were more likely to develop sensitivity. In spite of repeated requests, the data was not submitted.

The Friends of the Earth analysis concluded, "The StarLink debacle is a case study in the near total dependence of our regulatory agencies on the 'regulated' biotech and food industries. If industry chooses to submit faulty, unpublishable studies, it does so without consequence. If it should respond to an agency request with deficient data, it does so without reprimand or follow-up (e.g., statistics on allergic reactions reported to food companies). If a company finds it disadvantageous to characterize its product, then its properties remain uncertain or unknown. If a corporation chooses to ignore scientifically sound testing standards (e.g., by using surrogate protein without first establishing test substance equivalence), then faulty tests are conducted instead, and the results are considered legitimate. In the area of genetically engineered food regulation, the 'competent' agencies rarely if ever (know how to) conduct independent research to verify or supplement industry findings.

"One possible reason for this lapse is the FDA's avowed 'cheerleader' role in promoting biotechnology. Since a proper assay would more likely turn up an allergy 'problem', perhaps FDA chose the easy course of reliance on Aventis to avoid making trouble for the industry it openly promotes. This would be in keeping with the agency's history of subservience to the biotech and food industries with respect to genetically engineered foods."[23]

Some small amount of StarLink may linger in the human food chain forever. Although sold as a yellow feed corn, it has cross-pollinated into sweetcorn, popcorn and white corn, and was identified in the seed stock of 71 out of the 288 companies that the USDA contacted. We may never know if it was responsible for Grace Booth's anaphylactic shock or the symptoms described by countless others who called agencies and companies in the wake of the StarLink recall. Many more people may have been affected before that, without knowing the cause of their symptoms. For example, according to a letter submitted to the FDA in April 2000, five months before the American public knew anything about StarLink, one person "experienced immediate respiratory failure after ingesting two taco products."[23] He had a heart attack and died soon after. No one knew to question StarLink at that time. If it hadn't been for a privately funded effort by Friends of the Earth and others, we still might not know, and StarLink might still be on the market.

Other Bt Crops May Cause Allergies?
New evidence, on the other hand, reveals that allergies may also result from other varieties of the genetically engineered Bt crops still on the market. According to Hansen of the Consumers Union, "There is increasing evidence . . . that the various Bt endotoxins—including those from [GM corn], cotton and potatoes—may have adverse effects on the immune system and/or may be human allergens."[33]

In testimony before the EPA on October 20, 2000, Hansen described an EPA-funded study published in 1999 confirming that farm workers exposed to Bt insecticide sprays exhibited skin sensitization and the presence of IgE and IgG antibodies, both considered components of an allergic response. The workers with a greater reaction were those with more exposure to the spray—another allergy signal.[34]

While the workers did not exhibit respiratory symptoms, Hansen pointed out that the period of exposure was relatively short, and the amount of Bt that they were exposed to from the spray was quite small. Bt crops, on the other hand, have 10 to 100 times the amount of exposure. And the seeds of some of those Bt crops have yet another 10 to 100 times that amount. Thus, farm workers exposed to maize dust, for example, may breathe in up to 1,000 times the amount of Bt as those in the study. Those who work in mills and other processing plants would conceivably have an even greater risk.

"As part of the study," Hansen told the EPA, "the scientists were able to show that two of the farm workers studied had a positive skin-prick test." He pointed out that since these tests were now available to detect potential allergenicity of Bt crops, they should be immediately used to test people with high-level exposure to Bt proteins in both sprays and crops. The skin prick test didn't take long, was "relatively inexpensive", and "far more accurate than the . . . criteria presently being used" to evaluate allergenicity. Since this study was funded by the EPA, Hansen wondered "why the EPA hasn't already moved to conduct these tests."[35] The EPA didn't take Hansen's advice. They continue to rely on inferior criteria to assess the risk of allergens.

Three mouse studies were conducted on a Bt toxin, Cry1Ac, similar to that found in GM cotton and maize varieties. Two of these mouse studies showed that the Bt toxin triggers an antibody response in the blood and mucous membranes of mice; the third demonstrated that Cry1Ac boosts the immune response as powerfully as cholera toxin.[36] According to Joe Cummins, the mouse study and the farm worker study "clearly [show] that there is evidence that Bt crops are detrimental to mammals."[37]

More worrisome are the findings of a study published in *Natural Toxins*. The researchers were testing the effects of Bt potatoes on mice. As a control group, they spiked natural potatoes with a Bt protein that was not genetically modified. When the researchers analyzed tissue sections from the ileum—the lower part of the small intestine—they found a significant increase in cell growth—a potentially pre-cancerous condition. Although cancer in the ileum is rare, it empties into the colon, where cancer is common. According to Pusztai, if the Bt protein made it as far as the ileum, some probably made it into the colon as well; tests are needed to determine its effect.

According to the EPA, however, Bt toxin is not supposed to survive long enough to even get into the small intestine. It's supposed to be destroyed in the stomach. They base their claim on test tube experiments conducted by the biotech companies. The Bt protein is put into a test tube containing "simulated gastric fluid"—a mixture of hydrochloric acid and pepsin, a digestive enzyme—that crudely mimics the act of digestion in the stomach. The longer the protein stays intact, the greater the opportunity for it to elicit the antibody response associated with allergy.

Monsanto's test on its own Bt maize protein (Cry1Ab) resulted in over 90 percent degradation after just two minutes.[38] Critics point out, however, that the strength of the acid and the relative amounts of enzyme and Cry1Ab they put in their test tube was unrealistic and specifically designed to destroy the protein as quickly as possible.[39] Monsanto used a pH of 1.2, as compared to the considerably milder 2.0 recommended by the FAO/WHO.[40] And the ratio of pepsin to Cry1Ab used was about 1,250 times greater than the FAO/WHO international standard. In other words, Monsanto used a strongly acidic solution and a huge amount of enzyme to digest a very small amount of Cry1Ab, both greatly accelerating the rate of breakdown. When that same Cry1Ab was independently tested using the same conditions that StarLink's protein had been subjected to, 10 percent of Monsanto's Bt protein lasted one to two hours, not two minutes.[41] That's almost as tough as StarLink's Cry9C. If they had used the FAO/WHO guidelines, more of the protein would have lasted even longer. Furthermore, another test tube study showed that Cry1Ab only breaks down to sizeable fragments—large enough to remain potentially allergenic.[42]

Neither the EPA nor FDA has established standards for these tests—they have thus far accepted whatever procedures and conclusions the biotech companies have given them. Many scientists, however, criticize the EPA for even basing their claims on test tube studies. The critics say that these laboratory simulations are not reliable and insist that the longevity of proteins must be evaluated in the actual digestive systems of animals and humans. According to Pusztai, the surprising results of the mouse study supports this. He says, "Despite claims to the contrary, [the Bt] toxin was stable in the mouse gut."[13] This finding not only proves the FDA's testing methods are invalid, it undermines their important assumption that Bt breaks down too quickly to have any effect.

This begs an important question: Since the primary reason that StarLink was not approved was that Cry9C *might* survive digestion in the stomach, as soon as the Bt potato study confirmed that a natural Bt variety *did* survive past the stomach, wouldn't the EPA revoke approvals of the other Bts—or at least initiate an immediate investigation to verify the study's results? Even the scientists who made the surprise discovery concluded that it showed that "thorough tests" on GM crops were needed "to avoid the risks before marketing".[43]

The study was published in 1998. The EPA did not change the status of the approved Bts, did not initiate additional studies, and continues to rely on their test tube methods.

There's more. A second test for allergenicity involves comparing the structure of the foreign protein to those of known allergens. The reasoning is that if a section of the GM protein's amino acid sequence is similar to that of a known allergen, it might trigger a reaction. The EPA did not collect these important data for Bt maize before first approving it in the mid-1990s. Furthermore, it re-approved Bt maize in 2001 without demanding these data.

In 1998, an FDA researcher discovered a suspicious similarity between Cry1Ab and an egg yolk allergen. The study concluded that "the similarity . . . might be sufficient to warrant additional evaluation."[44] In 2002 Dutch scientists demonstrated that the two herbicide-resistance proteins used in Roundup Ready crops share sequences identical to those found in a shrimp allergen and a house dust mite allergen. The transgenic protein responsible for making GM papaya virus-resistant also possesses allergen look-alike sequences.[45] All of these data have thus far been ignored by regulators.

A third test for allergenicity is to see how well the protein survives heat treatment. This test not only indicates general stability, it also suggests how well the protein might survive food processing and end up intact in supermarket products. Here again, the EPA failed to collect the required heat stability data on Cry1Ab from the companies, and an independent study demonstrated that this Bt maize protein has "a relatively significant thermostability", comparable to the protein found in StarLink.

Based on these test methods, the most common varieties of Bt maize currently on the market would almost certainly fail the authoritative FAO/WHO testing protocol for allergenicity. When the EPA received the Friends of the Earth report, which detailed the shortcomings of the EPA's allergenicity review, the agency promised to respond. That was in 2001. There has been no response so far.

On February 22, 2004, the Norwegian Institute for Gene Ecology announced that at least thirty-nine people living adjacent to a large field of Bt maize in the Philippines were stricken with symptoms such as respiratory, intestinal and skin reactions, and fever, while the maize was pollinating. Although local authorities first suggested that the disease was infectious, this was contradicted when the symptoms of four families subsided after the members left the area, and then resurfaced when they moved back. Blood samples verified antibody responses to Bt-toxin, indicating an immune reaction to the pollen. The Institute announced its findings before the research had even been finished, because the alarming findings warranted immediate attention. The results are preliminary and the Bt maize is not conclusively linked to the symptoms.

Finally, there is one other finding that may destroy any remaining basis for the FDA's allergy safeguards. You may recall that Pusztai's potatoes grown from the same parent, with the same gene insertion, under identical growing conditions, had vastly different nutritional make-up. If one of those potatoes had been thoroughly tested for allergenic properties, the results would not necessarily be applicable for the others. Due to the shifting nutritional make-up of GM foods, accurate and reliable safety assessments *of any kind* may be impossible.

According to Pusztai, "The only thing you could do is find a stable GM organism, which has been put through tens of generations and still comes out the same, and which is not crossed with any other potato. You keep the purity of the line." But, he admits, this is not possible. He concludes, "We are storing up problems for the future."[9]

Missing Chickens

According to *BBC News*, April 27, 2002:

"Safety tests on genetically modified maize currently growing in Britain were flawed, it has emerged. The crop, T-25 GM maize, was tested in laboratory experiments on chickens. During the tests, twice as many chickens died when fed on T-25 GM maize, compared with those fed on conventional maize. This research was apparently overlooked when the crop was given marketing approval in 1996."[1]

Chapter 7

Muscling the Media

If you are learning about the many of the facts in this book for the first time, it is no accident. Many of the world's media, particularly in the United States, have been the target of an intensive pro-biotech campaign by the industry. Hence, there has been a chronic under-reporting of GM concerns—especially the health risks. The following stories provide examples of how public opinion about GM foods has been manipulated.

Muscling Television

When Monsanto's Bob Collier was asked why rbGH had not been approved in Europe he said the European Union "approved it technically from a safety standpoint, but the dairy policy there was such that they still have price supports . . . it proved to be a moratorium based on market issues not health issues."

Reporter Jane Akre, from a Fox television station in Tampa, Florida, was surprised at Collier's explanation. She thought, "I knew I had read a statement from the European Union stating that the as yet unknown health issues were a problem."[1] Akre was recalling a December 1994 letter from the Vice President of the Agriculture Committee of the European Commission to the director of the FDA stating, "Consumers in the European Community and their representatives in the European Parliament are apparently much more concerned about the unresolved human health issues related to recombinant Bovine Somatotropin than your agency was when it authorized the product."[2]

But Akre was a bit vague on the details and Collier, as Monsanto's dairy research director, was certainly an expert on the subject. Akre figured, "Oh well; he must know something I don't."

She asked Collier whether injections "rev" up the animal. He said the hormone "does not change the basal metabolic rate, it merely increases the amount of milk produced." Again Akre was surprised. She knew the drug had been called "crack for cows", and had read in Monsanto's own litera-

161

ture, "Cows injected with Posilac [Monsanto's brand name for rbGH] may experience periods of increased body temperature unrelated to illness." Akre thought, "Even the warning label that comes with Posilac makes some mention of an increased metabolism for the animal." But Collier was a senior fellow, a dairy scientist with a Ph.D. He was in charge of the division that sells rbGH. Akre once again decided that she must have been missing something. She would check her documents later.

Akre recalled, "Collier then told me that the cost of maintenance of an rbGH-injected animal doesn't change. 'That's not true,' I thought and asked him, 'What about higher feed costs and medical costs?' Collier replied, 'It does require more feed to produce more milk so I'm not saying you shouldn't provide more feed, it means there is no extra cost in addition to that for more milk.'"

"At this point," Akre said later, "I remembered the media-schooling that people from Monsanto go through before they appear on camera. And I think I just saw an example of the dance. I'm beginning to think it isn't me. In fact, I'm starting to get a little angry at being taken advantage of and at myself for letting that happen."

Akre redirected the conversation to IGF-1, the growth hormone associated with cancer. Akre recollected, "I asked about the limited testing for the effects of altered milk on humans. Collier tells me 'because the concentration of IGF-1 and bST doesn't change, there is no change in exposure, so the FDA concluded there is no indication that long-term chronic studies were justified.'"

Now Akre was ready. She reached into a stack of papers on her lap—research she had collected and some of the five pounds of documents sent to her by Monsanto, which, she is sure, they didn't expect her to read. Akre pulls out the Juskevich and Guyer report in *Science* 1990 that says Monsanto's own studies show an increase in IGF-1 in milk from treated cows. Collier responded by trying to reassure her that the National Institutes of Health (NIH) and the Government Accounting Office also review the process for human safety and concluded that the test process used by Monsanto was correct.

Again, Akre reached for her papers. She reported, "I pull out an American Medical Association report that says further study is needed as to the effects of IGF-1 on humans." Akre pointed out that the NIH also said more study is needed.

Collier then insisted that IGF-1 and bovine growth hormone (bGH) are digested, that there is no increase in concentration for bGH, and that bGH is not bioactive on humans. Akre interpreted this as a diversion to get her off the point. She knew that "dutiful reporters would write that down and the story would be over; no problem." But Akre refused to get sidetracked onto bGH. From what she read, IGF-1 is the real problem. And the studies show it is *not* digested.

Throughout the interview, Akre noticed that Collier fidgeted, cleared his throat, stumbled through his answers and was obviously uncomfortable. When Akre challenged him on an apparent contradiction, he would frequently say, as if rehearsed, "I'm glad you asked me that question." And Collier would habitually use the exact phrases that Akre would later hear repeated by other Monsanto and dairy industry spokesmen. "They all say exactly the same thing," recalled Akre. "It's the same wholesome product. . . . The milk is the same. . . . Our federal regulators have said consuming milk and meat from bGH-treated cows is safe. . . . It's not an issue to us or the FDA. . . . This is not something that knowledgeable people have concerns about."[1]

But Akre did have concerns. She and her husband, investigative reporter Steve Wilson, worked for three months, digging deep into broken promises, cancer links, corporate lies and influence in the FDA. Nothing was yet proven, but the red flags were there, especially concerning human health issues. All this and more was to be revealed to the public in a four-part news series. Or so they thought.

The reporters "seemed like a television dream team", reported the *Independent* in the UK. Akre was a former CNN anchorwoman and reporter. Wilson was a three-time Emmy Award winner whom *Penthouse* described as "one of the most famous and feared journalists in America" due to investigative reports he made exposing defects and hazards on Chrysler and Ford vehicles.[3]

WTVT Fox 13, a Florida TV station, hired Akre and Wilson in 1996 to beef up their news reporting. Within weeks, they were onto something. Wilson had discovered that although Florida grocers had publicly pledged not to buy milk from hormone-injected herds, they were doing so. And then in February 1997, Akre caught Collier on-camera making several statements that contradicted even Monsanto's own studies.

During his interview, Collier claimed that there wouldn't be any problem with increased levels of antibiotics in the milk since every truckload of milk is tested. But scientists and Florida dairy officials admit that each truckload is tested only for penicillin-related antibiotics. There's also a spot check for one other antibiotic done every three months. Their monitoring would miss the majority of the more than sixty varieties of antibiotics used by dairy farmers. Thus, it is likely that milk from rbGH-treated cows contains illegal levels and varieties of antibiotics. (According to data stolen from the FDA and published in *The Milkweed*, during a nine-month period from 1985 to 1986, employees at Monsanto's own experimental dairy farms applied more than 150 applications of a wide array of veterinary medicines not approved for dairy cattle by the FDA.[4])

Collier said on camera, "We have not opposed" voluntary labelling of products as rbGH-free. The reporters show, however, that Monsanto filed lawsuits against two small dairies to force them to stop labelling their milk as rbGH-free. According to *Rachel's Environment and Health Weekly*, "the dairies folded and Monsanto then sent letters around to other dairy organizations announcing the outcome of the two lawsuits—in all likelihood, for purposes of intimidation."[5]

Monsanto also supported legislation in Illinois that prevents dairies from telling consumers that their cows do not contain rbGH, and a research scientist reported that in spite of the New York City council vote of eleven to one to label milk with rbGH, "Monsanto was able to influence legislative votes so a mandatory label law was not enacted."[6] The documentary also reported that the Florida Commissioner of Agriculture and Consumer Affairs, who opposes labelling, received generous contributions from Monsanto toward his campaign and had been carefully schooled at a dairy conference on how to speak to consumers and discourage labelling.

When a Florida farmer informed Monsanto about health problems in his herd that started when he began rbGH, the farmer said that Monsanto told him "You're the only person having this problem so it must be what you're doing here, you must be having management problems." However, Monsanto had already found in its own research that "hundreds of other cows on other farms were also suffering hoof problems and mastitis, a painful infection of the cow's udders."[6] Furthermore, the law required Monsanto to notify the FDA about any adverse reactions such as the Florida farmer's complaints. But after four months of repeated phone calls by the farmer and even a visit by Monsanto to his farm, the FDA had heard nothing about it. Monsanto officials claim that "it took them four months to figure out that Knight [the farmer] was complaining about rbGH."[5] Knight eventually had to replace 75 percent of his herd.

The Fox news series even included an excerpt from Canadian national television in which a government official described how a Monsanto representative offered her committee a $1 to 2 million bribe if they recommended rbGH approval in Canada without further data or studies of the drug. A Monsanto spokesman said the officials misunderstood their company's offer of "research" funds.

The station invested thousands of dollars in radio advertising to promote the series, which was scheduled to air on Monday, February 24, 1997. But on the Friday before, Monsanto's lawyer faxed a letter to Roger Ailes, the head of Fox News in New York and the former director for media relations for President George H. W. Bush. The strongly worded letter detailed why the news series was, in Monsanto's opinion, biased and unscientific. In an interesting twist, one argument used in the letter was that "peer review is a basic protocol of scientific research."[6]

The letter also threatened, "There is a lot at stake in what is going on in Florida, not only for Monsanto, but also for Fox News and its owner." According to Akre and Wilson, this was the part of the letter that was of most concern to Ailes. Fox owned the Florida station; media mogul Rupert Murdoch owned Fox. Monsanto is a major advertiser with Fox TV nationwide. Moreover, Rupert Murdoch owns Actmedia, a major advertising agency used by Monsanto. If Monsanto pulled their advertising, this disagreement could be costly. The news series, which had already passed a review by attorneys, was pulled for "further review".

The Florida station's general manager, himself a former investigative reporter, did not back down. He studied the documentary with the station's lawyers and found that "nothing in the [Monsanto] letter raised any credible claim to the truthfulness, accuracy, or fairness of the reports."[6] He offered Monsanto another interview. Monsanto asked to see the questions in advance. The reporters insisted that no good journalist does that, but offered instead to supply a list of the topics. Monsanto declined the offer. The station re-scheduled the news series for a week later.

Monsanto's attorney immediately sent another, stronger letter to Ailes, this time indicating that the news story "could lead to serious damage to Monsanto and dire consequences for Fox News."[6] The airing was postponed indefinitely.

Soon after, the Florida station's general manager and news manager were fired. According to Wilson, the new general manager was a salesman with no experience in television. In one of their first meetings together,

Wilson realized the considerable gap between their motivations. To decide whether to run the story, the manager began by calculating the bottom line for the station. He figured he would lose advertising revenue from the supermarkets and from the dairy industry. Monsanto might also pull its advertising of agricultural products from Fox affiliates around the nation. Wilson tried to convince the manager to run the story on its merits. He said Monsanto's whole public relations campaign was based on the statement that milk from rbGH-treated cows is "the same safe wholesome product we've always known". But even Monsanto's own studies showed this to be a lie, and it could be endangering the public. Wilson recalled, "I tried to appeal to his basic sense of why this is news. He responded, 'Don't tell me what news is. We paid $2 billion for these television stations and the news is what we say it is. We'll tell you what the news is.'"[7]

According to Wilson, "He said, 'What would you do if I killed the story?' I said, 'I would be very disappointed.' But he asked me again what I would do. I couldn't figure where he was going with this. Again I told him I would be concerned. Then he made his point clear. He asked, 'Will you tell anybody?' Nobody had asked me that in all my years of television."

Wilson replied that he wouldn't go around beating a drum about it. But if someone asked why the show was cancelled, he wasn't going to lie. He told the manager, "I think I will refer them to you." At that point, the manager knew he had a problem. He couldn't count on Akre and Wilson to shut the whole thing up. So he embarked on a different strategy.

In a subsequent meeting the manager offered to pay about $150,000 to the couple. They would be paid the full amount of what was remaining on their contract, but they were free to go—essentially fired. But there was a catch. They were to agree never to talk about rbGH again—not on Fox and not for any other news organization.

But the veteran news team believed that "killing a public health story was unthinkable."[1] Wilson explained their position to the manager, "We think it's a matter the public should be aware of. We're not going to sell out our First Amendment rights to essentially do our jobs as journalists. I'm never going to agree for any amount of money you offer me to gag myself from revealing in some other time and place what's going on here."

Wilson said, "He looked at us with this blank stare like he'd never heard such a thing. And he said, 'I don't get it. What's with you people? I just want people who want to be on TV. . . . I've never met any people like you before.' He just offered us 6 figures and to him what we were being asked to do in exchange was no big deal. Why in the world would we turn it down? And lose a chance to continue to be on TV—as if that is such a big deal that one would sell one's soul to continue to do it."[7]

Thus, instead of giving in to his request for silence, they offered to re-write the documentary to make it more palatable. But each time they presented a script to Fox attorneys—who had taken over the editorial process—they were instructed to make it *more* favourable to Monsanto. Over the next six months, they did eighty-three re-writes.

Among the numerous changes, Akre and Wilson were instructed never to reveal that the FDA's approval of rbGH was based on "short-term" testing. They were allowed to include an interview with Samuel Epstein, M.D., who stated there are "lines of evidence showing that consumption of this milk poses risks of breast and colon cancer". The reporters were instructed, however, "not to include information that details the basis for this frightening claim."[6] They had to remove all mention of IGF-1 and any relevant studies and were not to use the word cancer again in any of the segments—referring only to "human health implications." The reporters also had to downplay Epstein's credentials. According to a website that documents the rewrites and the dispute, despite Epstein's "three medical degrees, a professorship of Occupational and Environmental Medicine at the University of Illinois School of Public Health, his frequent Congressional testimony as an expert on public health and environmental causes of cancer, his authorship of seven books [including the prize-winning 1978 book *The Politics of Cancer*] and countless editorials appearing in some of America's leading newspapers, [the] reporters were repeatedly blocked from describing him more completely. . . . Original references to him as a 'reputable scientist' which was acceptable in Versions 1–3, was later changed to 'respected scientist' which was acceptable in Version 11, and then 'well-credentialed M.D.' which was okay in Versions 10–18 until, ultimately, reporters were told no such reference was acceptable." The final reference was simply "Scientist, University of Illinois".[6]

Similarly, the credentials of a second scientist, William von Meyer, were stripped. The first version said: "Dr. von Meyer has spent thirty years studying chemical products and testing their effects on humans. He's supervised many such tests on thousands of animals at schools such as the University of London and UCLA. He's headed agricultural, chemical and genetic research at some of America's most prestigious companies." The final version of the script referred to him simply as 'scientist in Wisconsin'. The reporters were also ordered to remove his quote: "We're going to save some lives if we review this now."

Despite the intense scrutiny of every claim that opposed rbGH, Akre and Wilson "were repeatedly instructed to include unverified and even some outright false statements by Monsanto's dairy research director."[6] These included:

- Dr. von Meyer "has no credentials in human safety evaluation".
- "The cancer experts don't see the health issue. . ."
- "There are no human or animal safety issues that would prevent approval in Canada once they've completed their review, not that I'm aware of."

Monsanto's director also repeated a popular Monsanto claim that "Posilac [rbGH] is the single most-tested product in history." According to the reporters, however, "experts in the field of domestic animal science say that this claim is demonstrably false."

The journalists were told to leave in the Monsanto comment that "the milk has not changed"[6] as a result of injecting cows with the hormone. And they were eventually told to include a statement that milk from rbGH-injected cows is the same and as safe as milk from untreated cows. "Monsanto insisted that this statement be aired,"[3] said the *Independent*. According to the reporters, management even threatened to fire them if the statement was not included. But Akre and Wilson believed it wasn't true and presented scientific evidence to support their position.

Akre said, "We knew it was a lie. Monsanto's own study showed it was a lie. Yet we were told to leave that study in without refutation, even though we had contrary evidence. That's falsifying the news."[1]

After presenting all their evidence to Fox's lawyer demonstrating that Monsanto's claims were false, according to Wilson she replied, "You guys don't get it—it isn't about whether you have your facts right or whether it's true. It's the fact that we don't want to put up $200,000 to go up against Monsanto."[7]

Fox suspended the couple for "insubordination", then fired them altogether in December. Six months later, Fox hired another reporter, one with much less experience, to prepare another broadcast that contained the Monsanto statement.

Wilson said, "This is the first time I know of that a newspaper or broadcaster has opted not to kill a story but to mold the story into a shape that the potential litigant and advertiser would like." According to the *Independent*, "Fox categorically denies that it ever asked for false information to be included and says that the reporters were not willing to be objective."[3]

Akre and Wilson sued the station based on Florida whistle-blower laws. The jury awarded Akre $425,000. Fox appealed and the case was overturned. The appeals court ruled according to a strict interpretation of Florida law. Wilson and Akre had used the Federal Communications Commission's policy against news distortion as the basis of their claim. But this policy was not defined as a "rule, law, or regulation", required by the wording of the whistle-blower laws.

A few days after their decision, on Valentine's Day 2003, the court also declared that Akre and Wilson had to pay Fox's legal fees. Fox had hired more than a dozen lawyers, including former President Clinton's personal attorney, David Kendall. The costs are expected to run into the millions. The couple is planning an appeal to the Florida Supreme Court. They've given up trying to collect damages and are now just trying to protect themselves against the multimillion-dollar legal fees. According to Akre, "The ruling could gut all whistle-blower law in the state, if those who file complaints can be saddled with huge legal fees. You might as well throw the whistle-blower laws right out the window. It will kill them."[1]

The reporters have won several awards and recognitions, including a special award for Courage in Journalism from the Alliance for Democracy, the Joe A. Calloway Award for Civic Courage, and the Award for Ethics from the prestigious national Society of Professional Journalists. They were also the only journalists ever to receive the Goldman Environmental Prize, which included $125,000. Details of the suit and videos and scripts of the documentary are available at www.foxbghsuit.com.

Stifling Newspapers

On July 27, 1989, the *Los Angeles Times* published an op-ed piece on rbGH by Sam Epstein entitled, "Growth Hormones Would Endanger Milk".[8] Epstein outlined "grave consumer health risks that have not been investigated by the industry or FDA."

He wrote, "bGH and its digested products could be absorbed from milk into blood, particularly in infants, and produce hormonal and allergic effects." He described how "cell-stimulating growth factors . . . could induce premature growth and breast stimulation in infants, and possibly promote breast cancer in adults. . . . Also, the stress effects of the bovine growth hormones in cows could suppress immunity and activate latent viruses, such as bovine leukemia (leukosis) and bovine immunodeficiency viruses, which are related to the AIDS complex and may be infectious to humans." Epstein pointed out that the hormones in cows could promote the production of "steroids and adrenaline-type stressor chemicals . . . likely to contaminate milk and may be harmful, particularly to infants and young children." He said, "The fat and milk of cows are already contaminated with a wide range of carcinogenic contaminants, including dioxins and pesticides. Bovine growth hormones reduce body fat and are likely to mobilize these carcinogens into milk, with cancer risks to consumers." Epstein called for rbGH to be banned "until all safety questions can be resolved".

Soon after the letter appeared, senior Monsanto representatives visited the *L.A. Times* op-ed staff. They claimed that Epstein was scientifically unqualified and that the article was misleading. They urged the editors to turn down any future contributions from Epstein. The paper firmly rejected Monsanto's request.

Perhaps it was this meeting that convinced Monsanto to come up with a new strategy. Trying to defend their own product may have appeared too mercenary. Instead, they created what Epstein called a "hit squad" to do its bidding. Under the auspices of the public relations and lobby firm Capitoline/MS&L, they created a plan to identify and then stifle those reporters and reports that were critical of rbGH. They created a group called the Dairy Coalition, which included university researchers whose work was funded by Monsanto, selected "third party" experts and other organizations such as the International Food Information Council, which describes itself as "a non-profit organization that disseminates sound, scientific information on food safety and nutrition. . . ." According to the book *Trust Us We're Experts*, in actuality, the International Food Information Council "is a public relations arm of the food and beverage industries, which provide the bulk of its funding." Its past projects include defence of "monosodium glutamate, aspartame (NutraSweet), food dyes and olestra".[9]

In 1989, the coalition engaged the services of the PR firm of Carma International, which conducted a computer analysis of every news story on rbGH. Reporters were ranked as friends or enemies. The friends were rewarded; attempts were made to stifle enemies. Anyone who used Epstein as a source was an enemy.

Epstein had accumulated significant evidence about the potential health dangers of the hormone. In September 1989, he submitted his findings to the commissioner of the FDA, urging the agency not to approve the drug. His report, which went unanswered, outlined many of the key criticisms that the scientists at Health Canada were later to address. Epstein also received a box of secret FDA documents, sent to him anonymously. The information revealed that a high percentage of cows injected with rbGH had serious health problems and, according to Epstein, showed that Monsanto and the FDA were involved in a massive cover up.

In February 1996, the coalition tried to stop freelance writer Linda Weltner from including references to Epstein's concerns in her *Boston Globe* column. According to leaked internal documents from the Dairy Coalition, dairy officials wrote to the paper's assistant managing editor: "On [January] 23, Samuel Epstein . . . made unsupported allegations linking milk and cancer. . . . We're concerned that Ms. Weltner will give Epstein a forum in the *Boston Globe* to disseminate theories that have no basis in science." The letter claimed Epstein had "no standing among his peers in the scientific community and no credibility with the leading health organizations in this country." It said that "*USA Today* was the only newspaper to print these allegations and we recently held a heated meeting with them."[9]

At *USA Today*, coalition members had met with health reporter Anita Manning and her editor, after Manning had written an article that cited Epstein's concerns. The Coalition attacked Epstein's credentials. According to an internal Dairy Coalition document, "When Manning insisted it was her responsibility to tell both sides of the story, Callaway [of the coalition] said that was just a cop-out for not doing her homework. She was told that if she had attended the press conference, instead of writing the story from a press release, she would have learned that her peers from the *Washington Post*, the *New York Times*, *The Wall Street Journal* and the Associated Press chose not to do a story because of the source [Epstein]. At this point Manning left the meeting—her editor assured the Dairy Coalition that any future stories dealing with [rbGH] and health would be closely scrutinized."[9]

According to a February 1996 internal Dairy Coalition document, these other news sources didn't run a story because the Coalition had successfully educated the reporters. The same document says, "As you may recall, the Dairy Coalition worked hard with the *New York Times* last year to keep Marian Burros, a very anti-industry reporter, from 'breaking' Samuel Epstein's claim that milk from . . . supplemented cows causes breast and colon cancer. She did not do the story and now the *NYT* health reporters are the ones on the [rbGH] beat. They do not believe Epstein. Marian Burros is not happy about the situation."

The Dairy Coalition did not silence Epstein entirely. In another op-ed piece, he revealed, for example, that Congressman John Conyers, "chairman of the House Committee on Government Operations, requested Inspector General Richard Kusserow of the Department of Health and Human Services to immediately investigate the Food and Drug Administration for 'abdication of regulatory responsibility.' . . . Conyers charged that 'Monsanto and the FDA have chosen to suppress and manipulate animal health test data in efforts to approve commercial use of bGH.'"[10]

But coverage of rbGH was pretty thin. The Dairy Coalition had effectively kept critics like Epstein from getting their message into the mainstream media. This biased coverage continued when GM crops were introduced and impacted the op-ed and editorial pages as well.

This was confirmed by an April 2002 study conducted by Food First/Institute for Food and Development Policy. It revealed that "thirteen of the largest newspapers and magazines in the United States have all but shut out criticism of genetically modified (GM) food and crops from their opinion pages."[11] According to their press statement, their report "found an overwhelming bias in favour of GM foods not only on editorial pages, but also on op-ed pages, a forum usually reserved for a variety of opinions. In fact, the report found that some newspapers surveyed did not publish a single critical op-ed on GM foods and crops, while publishing several in support."

Anuradha Mittal, co-director of Food First/The Institute for Food and Development Policy, expressed concern that with such an important issue, differences of opinion "must be represented in the media if the public is to be able to exercise its democratic right to make informed decisions about new technologies."

The report showed that between September 1999 and August 2001, "Newspaper editorials were united in supporting GM foods and crops and only diverged on the issue of labelling." The arguments pitched in favour of GM foods were "by and large, the same arguments used by the biotechnology industry in their advertising campaigns." In the op-ed pages, "a forum usually reserved for a variety of opinions",[11] thirty-one out of forty pieces appearing in the major newspapers and magazines in America supported GM foods; only seven were critical. Another two argued for labelling.

These results may be due, in part, to the concentration of ownership in the US media. In the UK, where there is apparently more freedom to criticize GM foods, organizations like the Royal Society have tried to squelch that freedom. Not long after the British press's extensive coverage on Arpad Pusztai stirred up the public's distrust for GM food, the Society came up with a plan called "Guidance for Editors". They said it was to ensure that only "credible" scientists and research got into the press. According to the report "Suppressing Dissent in Science with GM Foods," "Before interviewing any scientist, the journalist will be expected to have consulted the officially nominated expert in the field." These approved experts would be listed in a directory published by the Society and would "be able to say whether the scientist in question holds correct views". Newspapers were not even supposed to publish opposing viewpoints to create a balanced story. Rather, the Society's approved advisors would establish the authenticity of the story, eliminating the need for minority viewpoints.

Not surprisingly, the pro-biotech UK government expressed support for the plan. The House of Lords Select Committee on Science and Technology even suggested additional restrictions of the press. According to their "Report on Science and Society", they wanted newspapers to avoid headlines that might damage the image of GM crops. Their second proposal, incredible as it may seem, attempted to purge the word "safe" from the vocabulary of the media. They suggest that "the very question, 'Is it safe?' is itself irresponsible, since it conveys the misleading impression that absolute safety is achievable."[12]

You Say Tomato—I Say "Not Any More"

"This tomato was picked seven days ago." The man held up a deep red tomato, about 2.5 inches in diameter, and showed it to the audience.

"This tomato was picked thirty days ago." Another deep red tomato was held up, about the same size.

"This tomato was picked sixty days ago." Now he had the attention of the nearly 500 attendees of the Minnesota Biotech Association. The tomato was identical.

"This tomato was picked ninety days ago. This tomato was picked 120 days ago. This tomato was picked 150 days ago," the man continued. He put all six tomatoes on the table. All six appeared fresh, red and ripe. All six had new genes in their DNA to keep them looking fresh.

The speaker paused, letting the room take in the miracle of this immortal tomato.

After some time, a man in his sixties stood up about twenty rows back. Everyone turned to listen as he broke the long silence. "As a biochemist, I have a problem. If this doesn't rot or decay in 150 days, then what have you done with the nutrient value." The man was Bill Lashmett. He had spent most of his life studying agriculture and ways to improve productivity, nutrition and soil ecology. And the 150-day tomato didn't fit into his definition of improvement.

The speaker at the front did not answer. According to Lashmett, two young men from the front row stood up in unison, walked back to where he was standing, and said in low tones that they didn't think that should be discussed there, but could he please join them outside?

The three went out of the hall, the door closed behind, and then one of the young men said, "We're not interested in the nutritive value. What we're interested in is if it's picked now, will a housewife buy it in 180 days."

Lashmett was incensed and told them so. He explained that if the tomato doesn't decay, then they must have done something with the sugars and enzymes. From a biological standpoint, that would be no good.

The two young men were polite, and let Lashmett continue talking for some time—probably happy that he was venting his anger out there with them, instead of back in the conference hall.

At the conference dinner that evening, Bill sat at a table alone with his wife. "You'd have thought I had leprosy," Lashmett said. "People would come up, look at us, and go somewhere else."[13]

Aware that his concerns were in the minority, Bill cancelled his membership in the Biotech Association and watched with sadness and anger at the rate that GM crops proliferated.

Pulping Magazines

"Food biotechnology is a matter of opinions. Monsanto believes you should hear all of them." That was the message in Monsanto's European-wide advertising campaign, designed to calm fears of GM foods.[14] As a direct response, the *Ecologist* magazine, "the established mouthpiece of the green movement"[15] in the UK, created a special Monsanto issue in 1998—The Monsanto Files—devoted to share some opinions of the biotech giant that would likely not appear in their ad campaigns.

According to an *Ecologist* press release, "The magazine highlights Monsanto's track record of social and ecological irresponsibility, and illustrates its readiness to intimidate and quash those ideas which conflict with its immediate interests."[14] An advance flyer about the issue read: "The giant Monsanto Corporation tells us that genetic engineering is all about feeding the hungry, about protecting the environment. But this is the company that brought us Agent Orange, PCBs and Bovine Growth Hormone: the same company that produces Roundup, the world's biggest selling pesticide, and the highly questionable 'Terminator Technology'. This Special Issue of the *Ecologist* asks the simple question: Can we allow corporations like Monsanto to gamble with the very future of life on Earth?"[15]

In September 1998, the issue was thought to have been mailed out as usual and the magazine's publishers were waiting in their office for the anticipated barrage of phone calls and media inquiries. But they waited and waited and no calls came. They soon found out why. Their printer, Penwell's of Liskeard, fearing a lawsuit by Monsanto, decided at the last minute not to send out the already printed magazine. Instead, they shredded all 14,000 copies.

The magazine's co-editor, Zac Goldsmith, said the *Ecologist* has "a long history of being forthright about issues and attacking powerful firms, yet not once in twenty-nine years has this printer expressed the slightest qualms about what we were doing."[16]

According to an *Ecologist* press release, the printer initially denied having had any contact with Monsanto. And "Monsanto's UK spokesman Daniel Verakis said he was mystified by the printer's action. 'The fact that the edition has been pulped is news to me. We had nothing to do with it," he said, adding that "he did not know that the edition was especially about Monsanto."[16]

But further discussions between the *Ecologist* and the printer revealed that the printer had spoken to Monsanto. They had contacted Monsanto, "seeking the assurance that any potential legal action would be taken against the magazine itself, and not against the small printing company. Their request was rejected," according to an *Ecologist* October 13 press release.[17] Without Monsanto's assurances, said Penwell's David Montgomery, "we weren't prepared to take the risk."[15]

When the *Ecologist* did find a printer two weeks later, their troubles were not over. Their October 26 press release announced, "Two leading newsagents in the UK, W H Smith and John Menzies, have recently confirmed that they will not be selling the controversial latest issue of the *Ecologist* magazine, for fear of being sued by the giant biotechnology company, Monsanto."

The frustrated editors said, "No one will deny the importance of balancing the one-sided messages put out by Monsanto in its advertisements and yet, in practice, it is almost impossible for critics to do so."

"Through reputation alone," said Zac Goldsmith, the magazine's co-editor, "Monsanto has been able, time and time again, to bring about what is in effect a de facto censorship. Their size and history of aggression has repeatedly brought an end to what is undeniably a legitimate and very important debate. They believe in information, but only that which ensures a favourable public response to their often dangerous products."[14]

Stop the Presses

In March 1998, Marc Lappé and Britt Bailey of the Center for Ethics and Toxics (CETOS) were anticipating the release of their book, *Against the Grain, Biotechnology and the Corporate Takeover of Your Food*. It was going to tell the world about the "perils of the genetic technologies in agriculture" and the corporate takeover of the food supply. The book took advantage of the combined experience of its authors. Lappé is an experimental pathologist and former director of the State of California's Hazard Evaluation System. This was his twelfth book. Bailey's graduate training is in environmental policy with a particular emphasis on the policies and regulations pertaining to new technologies.

But just three days before the book was to leave the printer, the publisher "received a threatening letter from the General Counsel's office of Monsanto Company". The letter referred to a short article that had appeared in *Coast Magazine* over five months earlier, which had excerpted sections from Lappé and Bailey's forthcoming 150-page book. Monsanto's attorneys claimed the article "was defamatory and potentially libellous against Roundup herbicide, Monsanto's major product."[18] The publisher, fearing the deep pockets of a hostile litigator, stopped the presses and cancelled the book.

The authors were outraged. The book had been reviewed by attorneys and was clearly not libellous. Moreover, Monsanto never bothered to request a transcript of the forthcoming book and waited until just three days before publication to protest, making any revisions before the press date impossible. Fortunately, the authors contacted the aptly named Common Courage Press, who did eventually publish the book about eight months later.

At the end of his letter, Monsanto's attorney went into some detail claiming that the phytoestrogen levels were no different between the Roundup Ready soybeans and conventional soybeans. Lappé and Bailey found this defence quite interesting, since they merely stated in the *Coast* article that no studies have been published examining the possible alterations in phytoestrogen levels. But the attorney's fervent defence of the issue tipped them off that there may be a problem, and they decided to investigate.

They examined Roundup Ready seeds and natural ones, careful to use isogenic varieties—meaning the two had the same parents, so to speak, the only difference being that the genetically modified variety also had Roundup Ready genes. The team discovered that compared to natural soya, the Roundup Ready varieties consistently had 12 to 14 percent fewer isoflavones—a type of phytoestrogen. In particular, the reductions were seen in the most biologically active isoflavones genistin and daidzin, both studied by the National Cancer Institutes and, according to Bailey, identified by nutritionists as being protective against heart disease, breast cancer, colon cancer, prostate cancer, and postmenopausal bone loss and osteoporosis.

Lappé and Bailey's discovery had serious implications. The health benefits of soya have been well publicized in recent years, in large part due to the presence of these phytoestrogens. The fact that GM soya, which comprises most US soya, might offer less protection from cancer, etc., might influence public health as well as public acceptance of Monsanto's beans.

The *Journal of Medicinal Food* agreed to publish the incriminating research in their July 1999 issue. In the meantime, Monsanto mounted a defence. They rallied the American Soybean Association (ASA), an organization that Monsanto had financially supported for years and which had become one of the staunchest biotech advocates. The ASA created a website that denounced the research and put out an article that attacked the findings. The article claimed that phytoestrogen levels normally vary widely and that the decreased amount was not significant. Levels vary, they said, due to differences in climate, temperature, soil and type of soybean. They failed to address the fact, however, that Lappé and Bailey used the same soybean variety grown under identical conditions.

To support their case, Monsanto produced a study of their own and published it in the *Journal of Agricultural and Food Chemistry* in November 1999. They reported that the phytoestrogen levels in their experiment varied so much, they couldn't even do a valid statistical analysis. When Lappé and Bailey investigated Monsanto's claim, however, they discovered how Monsanto had apparently designed the experiment in order to force this conclusion. In the research conducted by Lappé and Bailey, the extraction of phytoestrogens was done using the most up-to-date method. It had replaced an older extraction technique, which gave widely varied and less reliable results. When Monsanto researchers hired a lab to do their extraction, they instructed them to use this older, less reliable method. Sure enough, their results varied widely and they were able to defend their beans on that basis.

The reduced phytoestrogen levels that Lappé and Bailey found demonstrate a recurring problem with GM foods. Genetic engineering creates unpredictable changes; the composition of a GM food might be quite different from its natural counterpart.

Critics point out that the composition of Roundup Ready soybeans are significantly different from natural beans. In Monsanto's own study, the levels of ash, fat, carbohydrates, and trypsin inhibitor, a potential allergen, were all different. An investigator later discovered additional data had been omitted from the paper. It revealed that GM soya also had significantly lower levels of protein, a fatty acid, and an essential amino acid, as well as much higher levels of a potentially damaging lectin. The name that Monsanto chose for this study is telling: "The Composition of Glyphosate-tolerant [Roundup Ready] Soybean Seeds is Equivalent to That of Conventional Soybeans."

In addition to the conflict between the study's name and its own data, critics argue that Monsanto skewed the results by not spraying their beans with Roundup herbicide before testing nutrient levels. Herbicides can interact with plants, changing their chemistry. In the real world, the GM beans would always be sprayed before harvest. The whole purpose for growing Roundup Ready soybeans is so that farmers can spray their fields with Roundup herbicide, killing weeds but not killing the crop. In fact, studies show that farmers spray two to five times more herbicide on their GM soybeans compared to farmers who grow conventional beans.[19] The government actually increased the level of Roundup residue allowed on beans by three-fold to facilitate the sale of the herbicide-tolerant GM crops.[20]

In spite of the unsprayed beans and in spite of differences in nutrients, even if acknowledged by the scientific community, it would not likely inspire the FDA to remove GM soya from the market. That is, in part, because the concept of substantial equivalence is not defined. It is not a scientific concept; it is a subjective criterion. It is up to the whims and wisdom of the FDA regulators to determine what nutritional differences are allowed for GM foods. The FDA can also deflect any legal challenges on this point, since the court has ruled that their GM policy is not a rule but rather a non-binding guideline. Hence, the foundation of the FDA policy is a non-scientific, non-binding concept that allows GM foods into the market in spite of significant nutritional differences. Lappé and Bailey make this point in their newest book, *Engineering the Farm*, which did not get stopped at the printer.

The War over Butterflies

The researchers from Cornell University noted that the caterpillars, only three days old, were crawling more slowly than usual. They had been placed onto a milkweed plant that had been dusted with pollen from Bt maize. The Bt pesticide produced by the maize was supposed to kill the European corn borer. It was not supposed to affect these caterpillars. Nonetheless, according to a report in Bill Lambrecht's *Dinner at the New Gene Café*, by "the end of the four-day experiment, they fastened themselves in a death grip to the plant that constitutes their sole substance in life and their principal food when they become butterflies. Then they turned black and began to rot."[21] Forty-four percent of the caterpillars died. None of those exposed to non-GM maize pollen died. In May 1999, the prestigious journal *Nature* reported the study and all hell broke loose.

These were not just any caterpillars; they were destined to become monarch butterflies. Americans love monarchs. They're the Bambi of the insect world. And although the US press had failed to report virtually all evidence about the potential health risks of GM foods, an attack on monarchs was too much to ignore. The US press rallied to the butterfly's support and sent the biotech industry scrambling for an angle to defend themselves.

According to Lambrecht, "The first line of attack against the monarch butterfly study involved the usual nit-picking aimed at the scientific methodology used by Cornell's researchers. . . . The second line of attack involved rushing to sponsor a series of contrary studies." They set up a symposium to report the results of these studies, convened just six months after the *Nature* article had appeared. The meeting was sponsored by the industry-funded Biotechnology Stewardship Research Group.

The day before the symposium, the biotech industry set up a conference call with scientists and reporters, to announce the symposium's conclusions. Before the symposium started, the Biotechnology Industry Organization (BIO) put out a news release saying that "a panel of scientists is expected to conclude [that] genetically improved maize poses negligible harm to the monarch butterfly population." By the afternoon of the symposium, articles dated that day from the *Los Angeles Times*, the *Chicago Tribune*, the *St. Louis Post-Dispatch* and others were circulated among the scientists. According to Becky Goldburg of the Environmental Defense Fund, "The articles all reported that the symposium would conclude that Bt maize pollen posed little risk to monarchs—even though the articles had been written before the meeting had taken place!!" After the symposium, another news conference was set up with many of the same scientists that were part of the earlier conference call. The press conference was jointly sponsored by the biotech industry and the US Department of Agriculture and echoed the same conclusions.

Fortunately, the *New York Times* opted to send a reporter to actually cover the proceedings. According to Goldburg, "During the afternoon session, Carol Yoon of the *New York Times* stood up and said that she had just talked to her editors and that they had received a press release from industry stating that the symposium would conclude that Bt maize presented little risk to monarchs. Yoon asked if participants agreed with this conclusion. The answer was a clear 'No' from a number of researchers."

Goldburg said, "By the end of the day it became abundantly clear that the major purpose of the symposium, from the perspective of its sponsors, was not careful and deliberate evaluation of just completed, and in some cases, still incomplete scientific research. Instead, the meeting was designed and press interactions were orchestrated to provide the impression of scientific consensus when, in fact, no such consensus existed among meeting participants."[22]

In contrast to the articles that were written before the symposium by reporters who didn't attend, the *New York Times* article appeared with the headline, "No Consensus on the Effects of Engineering on Corn Crops".[23] Yoon wrote, "far from culminating in a consensus, the day was marked by sometimes heated exchange and ended with some scientists concluding that the bioengineered corn [maize] was safer than had been feared while others said that it was premature to draw any such conclusions. . . . Many of the researchers emphasized that their results were preliminary, with many studies still far from complete. . . . Some researchers expressed concern that so many studies, still far from completion and none peer-reviewed or published, should be given such a public airing, in particular in a forum orchestrated by the industry whose product safety has been brought into question. . . . 'We felt it was dirty pool and the fox was guarding the chicken coop,' said Dr. Lincoln Brower, a monarch expert at Sweet Briar College in Virginia. 'It was not conclusive.'"

It is interesting to note that when Arpad Pusztai spoke for just two and a half minutes about the conclusions of his research on GM potatoes, Monsanto's Colin Merritt complained, "You cannot go around releasing information of this kind unless it has been properly reviewed."[24] By contrast, Monsanto was one of the biotech sponsors of the butterfly symposium, in which some research was being reported after only 10 percent of it was complete and none of it had been reviewed.

A year after the symposium, another was held, also covered for the *New York Times* by Yoon. She reported that even though "corn and monarch butterflies are two of the best studied organisms on the planet," even with "a year and a half of research by more than twenty researchers from universities and industry, scientists . . . were still unable to say with any precision what the magnitude of risk was from the biotech corn to wild monarch populations." The cost to find that answer was estimated at "$2 million to $3 million, more than the Agriculture Department typically grants each year for the study of environmental risk," of GM crops. The head of BIO "said the public should not look to the private sector to foot the bill."[25]

The monarch butterfly taught the biotech sector a lesson. The damning research had hit them by surprise and they were forced to respond only after the media had alerted the public to the problem. Lambrecht describes one of their strategies designed to prevent that from happening again.

"I have a telling chart, compiled inside the biotech industry. It contains the web addresses of more than one hundred groups regarded as critics, the names of their members, details about the site registration, the groups' histories, and who links to whom. The industry devotes high-priced talent to monitoring anti-GMO activities, partly by using email addresses that do not divulge their companies' names. In the new century, industry strategists believed that they were monitoring enough sites to know when studies critical of their technology were about to surface. 'Now, we're able to pick these up well in advance and get our scientific evidence together so, unlike the monarch butterfly, they never get any traction,' an industry insider boasted."[26]

It turns out that the threat to the monarch butterfly did not catch everyone off guard. Arnold Foudin, an assistant director of scientific services at the US Department of Agriculture, said in an interview after the initial Cornell study was published, "We knew things like monarchs and other butterflies would be susceptible. That's part of the general background noise."[25]

Biotech Finds Its Poster Child

A national TV commercial showed a montage of smiling Asian children, caring doctors, rice paddies, and a narrator who says that golden rice can 'help prevent blindness and infection in millions of children' suffering from vitamin-A deficiency."[27] *Time* magazine went so far as to claim on their cover, "This rice could save a million kids a year." The biotech company Syngenta claims one month of a delay in marketing Golden Rice, would cause 50,000 children to go blind.[28]

The biotech industry had found its poster child, genetically engineered rice that makes its own beta-carotene—a precursor to vitamin A. In his *New York Times Magazine* article, "The Great Yellow Hype", Michael Pollan says that golden rice impales Americans on the horns of a moral dilemma: "If we don't get over our queasiness about eating genetically modified food, kids in the third world will go blind."

"Yet the more one learns about biotechnology's Great Yellow Hope," Pollan continues, "the more uncertain seems its promise."[27] A closer look reveals some interesting omissions in the industry's numbers. According to a Greenpeace report, golden rice provides so little vitamin A, "a two-year-old child would need to eat seven pounds per day."[29] Likewise, an adult would need to eat nearly twenty pounds to get the daily recommended dose.[28]

"This whole project is actually based on what can only be characterized as intentional deception," writes Benedikt Haerlin, former international coordinator of Greenpeace's genetic engineering campaign. "We recalculated their figures again and again. We just could not believe serious scientists and companies would do this."[30]

Even the president of the Rockefeller Foundation, which funded development of golden rice, said "the public-relations uses of golden rice have gone too far" and are misleading the public and media. He adds, "We do not consider golden rice the solution to the Vitamin A deficiency problem."[29]

"It remains to be seen whether golden rice will ever offer as much to malnourished children as it does to beleaguered biotech companies," says Pollan. "Its real achievement may be to win an argument rather than solve a public-health problem."[27]

There are other considerations as well. No published study has confirmed that the human body could actually convert the beta-carotene in golden rice. Also other nutrients such as fat and protein, often lacking in the diets of malnourished children, are needed in order to absorb Vitamin A. And it is not clear whether the genes from the daffodil, which are used to create golden rice, will transfer known allergens from the flower.[31]

The biotech proponents also admit that to persuade people to eat yellow rice may require an educational campaign. But if they are going to spend the time to educate, Pollan asks, why not instead teach "people how to grow green vegetables [that are rich in vitamin A and other nutrients] on the margins of their rice fields, and maybe even give them the seeds to do so? Or what about handing out vitamin-A supplements to children so severely malnourished their bodies can't metabolize beta-carotene?"[27]

Distributing supplements is precisely what the Vitamin Angel Alliance is doing. They give children who are at risk a high potency tablet, strong enough so that only two are required per year to prevent blindness. At a cost of only $.05 per tablet, only $25,000 is needed to prevent 500,000 children from going blind per year.[32] Contrast this with golden rice, which has cost more than $100 million dollars so far, and is not yet ready.

Michael Khoo of Greenpeace says golden rice "isn't about solving childhood blindness, it's about solving biotech's public relations problem." If the industry were truly dedicated to the problems of malnutrition and starvation, a tiny fraction of their advertising budget could have been diverted to make an enormous difference already. Khoo says, "It is shameful that the biotech industry is using starving children to promote a dubious product."[29]

Grains of Delusion, a research report jointly released by humanitarian organizations in Thailand, Cambodia, India, Philippines, Indonesia and Bangladesh, concluded that, "the main agenda for golden rice is not malnutrition but garnering greater support and acceptance for genetic engineering amongst the public, the scientific community and funding agencies. Given this reality, the promise of golden rice should be taken with a pinch of salt."[31]

Hiding the Food Safety Issue

Steve Druker had been aware that the US media was avoiding the GMO controversy, but he had just the story to change that. He had discovered that the GM policy of the FDA was against the law.

Druker, a public interest attorney, had read the laws over and over again and he was sure that the FDA had broken several. His organization, Alliance for Biointegrity, along with the International Center for Technology Assessment (CTA) in Washington, D.C., spearheaded a lawsuit to rein in the pro-biotech agency and force them to test GM foods and to label them. The suit had two lines of attack: religious and scientific.

On the religious front, Druker argued that by not labelling GM foods the FDA was not allowing individuals to practice their religious freedom. Based on three separate laws, the Food, Drug and Cosmetic Act, the US constitution, and the Religious Freedom Restoration Act, Druker reasoned that individuals who were religiously opposed to eating GM foods must be able to identify them in order to avoid them. Several clergy and religious organizations became plaintiffs in the case, including seven Christian clergy, three rabbis, a prominent Buddhist and a Hindu organization. These plaintiffs from diverse religious backgrounds all viewed genetic reconfiguration of foods as a violation of the basic principles of their faith. They felt obliged to avoid these foods and live in accord with their beliefs, but were hampered by the inability to identify which foods were GMOs.

On the scientific side, Druker believed that the FDA had violated the law by presuming that all GM foods are Generally Recognized as Safe (GRAS). This was a critical presumption. The FDA claimed that because these foods are GRAS, they do not need to be tested for safety. But there are strict criteria for GRAS status, and one of them entails testing. The criteria are: 1. There must be a scientific consensus that the food is safe. 2. The consensus must be based on empirical evidence demonstrating safety, and such evidence should ordinarily be published in peer-reviewed journals. In the past, even a few well-credentialed scientists who did not believe that a product's safety had been established was enough to prevent it from being listed as GRAS.

GM Foods did not meet either criterion. There were no peer-reviewed articles demonstrating the safety of any of these foods, and many eminent scientists believed GM foods to be unsafe. Hence, by claiming they are GRAS, the FDA had apparently violated the law.

To illustrate to the court that there wasn't consensus among the scientific community, Druker put together an impressive group of nine scientists who challenged the FDA's safety claims. Furthermore, these scientists actually joined the lawsuit as plaintiffs. This was unprecedented. While scientists regularly take the role of advisors or expert witnesses, these nine were suing the FDA.

Two members of the team worked in the field of biotechnology, although not with GM food. From their own work, they were familiar with the risks associated with transferring genes across species and were quite concerned that this imprecise technology was being applied to food—risking the health of the population. Also among the plaintiffs were Professor Philip Regal, a renowned expert in plant genetics from the University of Minnesota and Richard Strohman, professor of molecular and cell biology at UC Berkeley. Providing supporting testimony was food safety expert Richard Lacey, M.D., Ph.D., the first scientist to publish warnings about the threat of Mad Cow disease.

The very fact that prominent scientists were suing the FDA and publicly declaring that GM foods cannot be presumed safe should, Druker reasoned, demonstrate that there was not a consensus on safety. The FDA's claim of GRAS would therefore be clearly discredited.

On May 28, 1998, the day the suit was filed in US District court in Washington, D.C., religious leaders and scientists convened at a press conference at the National Press Club, an event that was sure to make the headlines. But instead of informing the public about the lack of consensus among scientists, the absence of peer-reviewed research, or how the FDA violated the law by not requiring safety tests, the media reports of the event focused primarily on the religious issue and other aspects of the labelling debate. Most of the discussion of scientific and safety issues came from the FDA and biotech representatives, who assured the public that the foods were proven safe. Even the fact that eminent scientists were plaintiffs in the lawsuit was overlooked by most of the coverage. Furthermore, the *Washington Post*, *New York Times* and *The Wall Street Journal* did not report on the lawsuit at all. Druker got a first-hand look at the bias of the US media. But it didn't stop there.

During the lawsuit, the FDA was required to give the plaintiffs' attorneys more than 44,000 pages of its internal files. After sorting through the mountain of documents, Druker and the other attorneys discovered clear evidence of fraud and cover-up. The FDA policy had claimed that the agency was not aware of any evidence that GM foods differed from normal, natural foods in any meaningful way. But memo after memo from the FDA's own scientists revealed just the opposite. There were concerns about toxins, allergens, new diseases, nutritional differences and environmental dangers, and there were unresolved issues about the feeding studies on the FlavrSavr tomato. These documents were the smoking guns. They proved that there was not a scientific consensus and that the law had been broken.

With evidence in hand, Druker and others spoke at a well-publicized press conference in Washington, D.C. in June 1999. Druker was later interviewed by reporters from the *Washington Post*, *New York Times* and *The Wall Street Journal*. But none of the resulting stories mentioned the FDA lies and cover-up.

The coverage by *The Wall Street Journal* focused exclusively on the religious angle and, according to *Salon Magazine*, portrayed Druker as "something of a small-town, Torah-thumping fanatic".[33] For example, the *Journal* article entitled "Motley Group Pushes for FDA Labels on Biofoods", reported that Druker "began crisscrossing the country, gathering his Noah's Ark of plaintiffs, many of whom share his mystical spirituality and distrust of authority."[34]

The only mention of the lawsuit from the *Washington Post* came in the middle of an August 1999 article on GM food labelling issue. It said, "Last summer, two consumer groups sued the Food and Drug Administration, claiming that the agency's failure to institute a labelling regimen for gene-altered food is in violation of the Food, Drug and Cosmetic Act. The law demands that food additives not 'generally recognized as safe' be labelled."[35] The article also included the standard quotes about safety from the Biotech Industry Organization.

Druker says, "It was like there was a decision made in the media that they'll talk a little about the environment, but they would not report that there are scientific grounds for concerns about food safety."

The *New York Times* appeared to be interested in Druker, interviewing him several times over the coming months. Finally, in January 2001, a year and a half after Druker's press conference, the *Times* ran an in-depth story on the history of Monsanto's influence at the FDA, which did include quotes taken from a couple of FDA scientists warning their superiors about the health risks of GM foods. The article was unprecedented, giving American readers an insight into government corruption surrounding approval of GM foods.

But Druker scored few victories like this in the US press. Their pro-biotech bias would haunt him for the next four years. He would be interviewed and stories would be written, only to be cancelled by the editor. If anything did make it to print, it was trivialized and played down. Rarely was there ever a mention of scientific concerns for human safety.

One telling incident occurred in August 1999, when an ABC national news producer called Druker from Washington and asked him to drive ninety minutes to their affiliate's studio for an interview about his lawsuit. By this time, Druker was savvy. He told the producer that he was tired of spending lots of time and energy to get an interview only to have it canned before production. He laid down his requirements. He would grant the interview only if they reported on the concerns about GM food safety that the FDA's scientists had raised and that the FDA had ignored. The producer, according to Druker, agreed to be fair and balanced in the reporting. Druker made the drive.

ABC news flew in a crew. The interview lasted 15 minutes. A couple of days later, ABC news aired a three-minute story about GM foods. The primary scientist featured was BIO's president, Michael Phillips. Druker was on for less than a sentence. Which sentence? Naming some of the religious denominations involved in the suit. The points about the FDA scientists were not mentioned. According to Druker, the coverage was designed so that "people watching would have no legitimate reasons for concern."

Druker has spoken about the FDA and the details of his suit on five continents; in every other country, the FDA's cover-up is extensively reported. In the US, "it's as if there is an implicit agenda to suppress it,"[36] says Druker.

On October 2, 2000, the federal court issued their ruling in favour of the FDA on technical grounds. According to Druker, "The court determined that the FDA is **not** regulating GE [genetically engineered] foods at all. . . . It declared that the FDA's policy on GE foods is essentially one of 'inaction' and does 'not impose any . . . obligations' on the biotech industry." Since they had done nothing to regulate the industry either before or after their GM food policy was issued, the entire grounds for his case was overruled. Druker said the court acknowledged that "The FDA's politically appointed bureaucrats did not follow the advice and warnings of the agency's scientific staff regarding GE foods but disregarded them, [and] there is currently significant disagreement among scientific experts about the safety of GE foods." Druker continued, "Further, the court avoided the issue of whether adequate safety testing has been done and failed to make a determination that GE foods have been demonstrated to be safe— even though such a determination is legally required in order for these foods to be on the market."

Although the decision was appealed, in January 2001 the FDA proposed new regulations, which forced Druker to withdraw the appeal and then wait to re-introduce a new lawsuit after the new laws go into effect.

Although Druker didn't win his case, he says, "our lawsuit accomplished a lot by exposing the FDA's fraud and revealing the unsoundness of its policy and the irresponsibility of its behaviour. Even though we failed to overturn the FDA's policy, the court's ruling refutes the standard claims of the biotech industry about the rigor of FDA oversight and the proven safety of its own products. It gives the FDA nothing to be proud of nor does it give the biotech industry anything to brag about. But it does give all consumers something to be very concerned about."[37]

Thwarting Consumer Choice

Curious about consumer response to GM foods, in 2000 the FDA conducted twelve focus groups around the country where they interviewed citizens about the issue. It turned out that most people didn't know they were eating GM foods, let alone eating them at almost every meal. When they found out, many were outraged. Virtually everyone said they wanted the food to be labelled. They were concerned about long-term health effects and wanted to have the choice whether to eat GM foods.

The desire for labelling was not a surprise. Every independent poll has confirmed that citizens around the world want GM foods labelled. Various polls in the US show that 70 to 94 percent of the population favour mandatory labelling of GM foods. Almost all industrialized nations have responded to consumer desires and now require labelling—but not the US.

The stated policy of the United States is to promote GM foods, and many believe that labelling would hamper that goal. In fact, a *Time* magazine poll confirmed that 58 percent of Americans said that if GM foods were labelled, they would avoid purchasing them.[38] labelling is therefore not part of the government's agenda, regardless of citizens' desires.

Many have challenged the US position. Laura Ticciati, founder of Mothers for Natural Law and co-author of *Genetically Engineered Foods: Are They Safe? You Decide*, delivered nearly 500,000 signatures to the nation's leaders on June 17, 1999, asking that GM foods be labelled. Ticciati says, "Despite the clear message that the American people want to know what's happening to their food, our government just continues serving the interests of industry rather than the rights of its people. It's completely indefensible to tell mothers they don't have the right to know what's in their children's food."[39]

Congressman Dennis Kucinich said of labelling, "There's something very American about it. People want the right to know. We're the country of freedom of information."[40] He introduced labelling legislation in the House in 1999, and Senator Barbara Boxer tried the same in the Senate, but the bills never came up for a vote.

A handful of citizens from Oregon decided to take matters into their own hands. Taking advantage of voter laws in their state, they collected over 100,000 signatures on their petition and placed a labelling bill on the ballot in November 2002. Measure 27, as it was designated, would have required any food containing an ingredient with more than 0.1 percent GM content to be labelled. Further, if GM processing agents, hormones, or anything related to genetic engineering were used in the food's production, it would also need a label.

When the measure was first introduced, nearly 60 percent of Oregonians polled were in favour. But then the biotech industry moved in. They spent $5.4 million—twenty-five times the amount spent by pro-27 campaigners—telling Oregon voters to vote no. In the end, the measure was defeated. Only 30 percent voted in favour.

How did the biotech industry convince people to vote against what citizens everywhere else have consistently been in favour of? According to Craig Winters, Director of the Campaign to Label Genetically Engineered Foods, they used fear and distortion. For example, in an eight-page brochure mailed throughout the state, the biotech industry gave a chart claiming the average grocery bill would skyrocket by $550 per year. They backed up these figures with the reference: "Economic Analysis of Oregon Measure 27, August 30, 2002." According to Winter it was an industry-sponsored, indefensible study designed to give inflated results.

A more reliable analysis, says Winter, was conducted by William Jaeger, an economist and agricultural policy specialist at Oregon State University (OSU). Looking at analyses used in other countries, the estimated costs for labelling ranged from $.23 to about $10 per person per year. One study that "was based on more limited information and a less detailed analysis of the costs"[41] put the annual per person figure at $35 to $48. Even this higher estimate is in sharp contrast to the $550 per family that the Oregon voters read about.

In addition to the brochure, voters were bombarded with TV and radio ads repeating the exaggerated figures, but failing to even mention genetic engineering much of the time. The media campaign also claimed that Measure 27 would hurt farmers, restaurants, businesses, the government and regular citizens. Winter says, "If you tell a lie enough times, people believe it."

Monsanto contributed $1.5 million to the Oregon campaign to prevent labelling. Contrast that with the interesting spin in Monsanto's European ad in June 1998, which read: "You have the right to know what you eat, especially when it's better. . . . After several months of debate, Europe has just adopted a new law for the labelling of food that comes from genetically engineered plants. . . . We believe that products that come from biotechnology are better and that they should be labelled."[42]

In addition to support from biotech and food companies, the anti-Measure 27 campaign got support from the FDA. In an unprecedented move, Lester Crawford, deputy commissioner of the FDA and a former Executive Vice President of the National Food Processors Association, sent a letter to the office of Oregon Governor John Kitzhaber, strongly objecting to the measure. The letter, which was reproduced in the industry's brochure to voters, said, "FDA is not aware of any information or data that would suggest that any genetically engineered foods that have been allowed for human use are not as safe as conventional foods."

Incensed that the agency was repeating the same line even after internal documents exposed it as a lie, attorney Steve Druker wrote to the Governor citing several quotes by FDA scientists that proclaimed just the opposite.

Druker also wrote, "Dr. Crawford's letter further misrepresents the facts by stating: 'FDA's scientific evaluation of bioengineered foods continues to show that these foods . . . are as safe as their conventional counterparts.' This claim is quite curious in light of the agency's statement reported in the *Lancet*, May 29, 1999: 'FDA has not found it necessary to conduct comprehensive scientific reviews of foods derived from bioengineered plants—consistent with its 1992 policy.' Since the FDA requires no testing of GE foods, acknowledges it does not conduct comprehensive reviews of them, and does not make formal empirical findings that particular GE foods are safe, it's amazing the agency would now claim its evaluation process shows they are as safe as other foods."[43]

It is telling that in a speech before the International Dairy Foods Association in January 2003, Crawford described the goals of the FDA as ensuring food safety and promoting the development of biotechnology.

One of the organizations that worked hard to stop the measure was Oregonians for Food and Shelter. Although this beneficent sounding group claimed to run a grassroots campaign, board members include Monsanto and Dupont, and the group's stated goal is to promote "pest management products, soil nutrients and biotechnology". In their mailing to voters, they repeated the inflated $550 figure and claimed that Measure 27 would add a whopping 32 to 63 percent to the cost of family farms and food processing plants. The letter said, "Measure 27 is another example of narrow special interests trying to use Oregon's ballot measure process to push their extreme political agenda."[44]

In April 2003, the same organization is pushing a bill through the Oregon legislature that "would keep local governments from imposing any food labelling requirements and would prevent state agencies from adopting requirements stricter than the federal government allows." Richard North, project director of the Campaign For Safe Food, sees the bill as a way to hinder future citizen initiatives like Measure 27 and legislative action. "What's getting clobbered here is the consumer's right to know,"[45] he said. The efforts to prevent future labelling in Oregon are reminiscent of the so-called Food Defamation laws passed by thirteen states. According to the *Guardian*, these laws, which came about due to heavy lobbying by the biotech industry prevent the "spreading of false and damaging information about food."[46]

Pressuring Scientific Opinion

In late September 2001, Ignacio Chapela, a microbial ecologist from the University of California at Berkeley, rode in the taxi with an official from the Mexican government. The official had waited all day for Chapela to finish his meetings, so that he could escort the scientist to see Fernando Ortiz Monasterio, Mexico's top man in charge of biotechnology. They drove through a rough part of Mexico City to a government office building. It was early evening and Chapela was surprised that no one was around. He was taken up to the twelfth floor, escorted down a dark corridor and led into a scene that was to leave him shaken hours later.

Everything appeared to be arranged for maximum effect—maximum intimidation. Monasterio, the Director of the Commission of Biosafety and GMOs, sat behind a makeshift desk—the door to the office had been taken off and laid across cardboard boxes. He welcomed Chapela coldly. After a maid had poured coffee, she was asked to leave. Chapela sat before the director, and Monasterio's assistant sat down behind Chapela blocking the doorway. There was no other office furniture.

Monasterio glared at Chapela and proceeded to recite what appeared to be a well-thought out attack that lasted more than an hour. "First he trashed me," said Chapela. "He let me know how damaging to the country and how problematic my information was to be. He said, 'You are about to create a problem. . . . We are looking forward to the day when these technologies are going to come to our country, but there is a hurdle and that hurdle is you.'"[47]

At one point Chapela was shown around the offices, a tactic that he believed was to increase his apprehension. He says that no one else was present, and surrounding the building was nothing but dump sites. The only phone was the director's cell phone. At one point, Chapela nervously laughed and said, "Are you going to take a gun out and shoot me." He appeared to joke, but he was scared. Monasterio wasn't comforting. According to Chapela, Monasterio wanted him to withdraw from publication the incriminating evidence that Chapela and Berkeley Ph.D. student David Quist had uncovered in their research.

Mexico is home to hundreds of indigenous varieties of maize, which crossbreed naturally to create strains that are most hardy for the region.

"To preserve this gene bank," reported the *Guardian*, the Mexican government "banned planting of GM crops in 1998".[48] They feared that cross-pollination might contaminate the indigenous maize species. Such contamination would be permanent—there is no known way to clean up the gene pool. Not everyone in the Mexican government was happy about this ban. Officials like Monasterio wanted their country to embrace biotechnology and wanted to allay concerns about gene contamination.

In spite of the ban, Mexico imports maize from the US for food purposes and some of the maize is used for planting instead—illegally. Because about 30 percent of US maize is genetically modified, about the same proportion of the maize grown from this seed is also modified. Quist and Chapela tested indigenous maize in more than a dozen communities in the remote mountain region of Oaxaca and, to their surprise, discovered that 6 percent of the plants tested had been contaminated with GM maize. If contamination had penetrated this far, it was sure to be widespread. The prestigious journal *Nature* had agreed to publish this controversial finding, which threatened to disrupt the biotech industry's attempts to convince Mexico, Brazil and the European Union to go forward with planting GM crops. And now, according to Chapela, Monasterio was telling him to stop publication.

Chapela did not give in. The few times he was able to respond, he tried to explain that he was not the cause of the problem—he had just discovered it. Furthermore, the ministry of agriculture was already verifying his findings.

After about an hour, Monasterio changed his tactics. According to Chapela, the director said, "You have created a problem and I will give you an opportunity to be part of a solution. I am going to run the research that is going to show the world what the truth is." Monasterio said that a team of five of the top molecular biologists were going to do the research and discover that Chapela's study was not correct at all. Furthermore, Chapela was invited to be part of this team. The others were to include two scientists from Monsanto and two from DuPont. The plan was for the five of them to meet at one of the top private resorts in Mexico and have their research completed in just six weeks. Furthermore, they would submit their work to *Nature*. Chapela was told, "So don't worry, you will get a *Nature* publication." Chapela explained that he was a professor at UC Berkeley—a public institution—and could not work on a private research project.

Chapela was feeling quite regretful that he had informed the Mexican government of his research. He had told them, as a matter of courtesy, so they could be prepared to respond when the news became public. He also told them of *Nature*'s strict exclusivity requirements: if the research became public before they published it, the paper would be withdrawn. He therefore had asked them to keep the findings confidential. Chapela said to the director, "You may derail the *Nature* publication by bringing it to the media, but you are not going to stop me from trying to get it published."

Monasterio appeared to give up. He walked Chapela out of the building and then insisted that he personally give him a ride in his SUV back to his hotel. According to Chapela, "He started asking personal questions about my kids. He asked where my daughter goes to school. And he wanted to take me precisely to the place where I was staying." Before letting him out of the car, however, Chapela says that Monasterio implied, "Now we know where your children go to school." Chapela later told the BBC, "I was emotionally very shocked and drained. I felt totally shaken, and I just stayed in a state of shock for hours afterwards."[49]

The day after the meeting, Monasterio called a meeting with Greenpeace and others to announce Chapela's findings. Greenpeace was not willing to wait the two and a half months until publication to start their campaign and told Chapela that they would have to bring his findings to the media. Thus, according to Chapela, Monasterio had indirectly leaked the research to the press in an attempt to violate *Nature*'s rules about pre-publication secrecy.

In an interview with the BBC, Monasterio acknowledged that he met with Chapela but denies allegations that he was threatening. He also said the meeting took place on the fifth floor in the Ministry's office.[49]

Nature was not dissuaded by the advanced publicity and kept the November 29, 2001 publication date. A day or two before, Chapela received a fax from Victor Villalobos, the under-minister for agriculture and a close colleague of Monasterio. Chapela says that while the details of his meeting with Monasterio were undocumented and unconfirmable, the fax from Villalobos was hard evidence of the government's attempt to suppress the information through intimidation. And the fax expressed the exact tone that Monasterio had used in person. According to Chapela, the fax claimed that the government is the only legitimate body to conduct that type of research. Further, Chapela would be held personally responsible for all damages caused to agriculture and to the economy in general, by his publication. The government would take all necessary actions to redress the situation.

The day the paper was published, messages started to circulate on a biotechnology listserve called AgBioWorld, which is distributed to more than 3,000 scientists. The first message came from a Mary Murphy, which charged that Chapela was biased. Then, came a message from Andura Smetacek, falsely claiming that Chapela's paper had not been peer-reviewed. The message also accused Chapela of being "first and foremost an activist" and said his research was published in collusion with environmentalists. Chapela couldn't immediately respond, since his internet service shut off for three days, just at the time of the postings. Some suspect a well-timed hacker attack.

Smetacek followed the next day with another attack on Chapela's credibility. The wording on these messages was powerfully written and soon hundreds of other messages appeared, repeating or embellishing the accusations. The listserver, AgBioWorld, launched a petition to be sent to *Nature*. Scientists on the email list were eager by then to sign it and *Nature* was besieged by a worldwide campaign asking that they retract the article.

The arguments against the paper were focused not on the discovery of contamination, but on a second conclusion made in the article, which had even more serious implications. According to their tests, Quist and Chapela discovered that the contaminated GM maize contained as many as eight fragments of the CaMV promoter. There could be several reasons for this. The one cited by GMO critics was that this discovery demonstrated that the CaMV promoter created an unstable "hotspot", described in Chapter Two. They believed that when the pollen contaminated the native plants, the hotspot caused the genes to fragment and promiscuously scatter throughout the plant's genome. If such genetic instability were verified, the impact on a plant species would be devastating. Moreover, any pretense of safety, precision, or predictability of the effects of GM crops would have to be abandoned and GM foods would likely be finished.

While the evidence that GM maize contaminated local indigenous varieties was solid and easily verified, the second conclusion was not as well established. Even the authors admitted in the article that the testing procedure they used to identify where along the DNA the eight fragments of the CaMV promoters were located was based on an exploratory method and open to interpretation. The hotspot hypothesis could not be verified. The pro-biotech scientists, however, claimed that the article was promoting the hotspot hypothesis without adequate proof and blasted away at the paper's credibility.

As a result of the enormous pressure placed on it, *Nature* did something unprecedented in its 133-year history. The editor wrote a partial retraction, suggesting that the conclusions surrounding the extra promoters were not sufficiently supported. *Nature* upheld the central finding— that GM maize had contaminated natural maize stock in Mexico.

This distinction was lost, however, on the world's media. The *London Times*, for example, incorrectly reported that "*Nature*, one of the world's most prestigious peer-reviewed journals, admitted yesterday it had been wrong to publish flawed research that claimed to prove that genes from GM maize had accidentally crossed into a traditional variety in Mexico."[50] An Associated Press report said, "Encouraged by PR firms working for Monsanto and other companies," the media "reported *Nature's* retraction as a 'big public relations victory for the biotechnology industry'[51] and as one pro-GE scientist stated, a 'testament to the technical incompetence' of biotech critics."[52]

Upon closer examination, however, it became apparent that the PR victory was not at all spontaneous. Both Mary Murphy and Andura Smetacek, who started the email campaign attacking the research, claimed in emails to be ordinary citizens, devoid of corporate links. According to columnist George Monbiot of the *Guardian*, they are in fact fabricated names being used by the Bivings Group, a PR firm that works for Monsanto. Monbiot alleges that "Mary Murphy is being posted by a Bivings web designer, writing from both the office and his home computer in Hyattsville, Maryland; while Andura Smetacek appears to be the company's chief internet marketer." The head of online PR did eventually admit to the BBC's *Newsnight* that one of the emails was sent from someone "working for Bivings" or "clients using our services".[53]

In addition to rallying scientists to challenge Chapela's article, these internet pseudo-humans have been busy stirring up pro-biotech sentiment for some time. Starting in 2000, Andura Smetacek repeatedly accused GM critics of terrorism. One of Smetacek's letters, which accused Greenpeace of deliberately spreading fears about GM foods to further its own financial interests, appeared in the *Glasgow Herald*. Greenpeace sued the paper for libel and won. A closer look at three of Smetacek's emails, including the first one sent, had an internet protocol address assigned to the server gatekeeper2.monsanto.com—owned by Monsanto.

Investigators have also linked the website of AgBioWorld—the organization that arranged the listserve and the *Nature* petition—to Bivings, as well as the Center for Food and Agricultural Research. According to Monbiot, "The center appears not to exist, except as a website, which repeatedly accuses greens of plotting violence."[53]

The Bivings Group specializes in internet lobbying. On their website is an article entitled "Viral Marketing: How to Infect the World". Although it appears that the content of the article might have changed since his article was published, Monbiot quotes excerpts as follows: "There are some campaigns where it would be undesirable or even disastrous to let the audience know that your organization is directly involved . . . it is possible to make postings to these outlets that present your position as an uninvolved third party. . . . Perhaps the greatest advantage of viral marketing is that your message is placed into a context where it is more likely to be considered seriously."

Monbiot quotes another section of the website: "Sometimes we win awards. Sometimes only the client knows the precise role we played." On the front page of their website (as of October 21, 2002) is an announcement about an award they recently received—for work done for Monsanto.

Monsanto's director of internet outreach, Jay Byrne, "told the internet newsletter Wow that he 'spends his time and effort participating' in web discussions about biotech. He singled out the site AgBioWorld, where he 'ensures his company gets proper play.'" Another method Byrne used was to manipulate websites so that search engines listed only pro-biotech sites at the top of their list. Many of these sites were virtual organizations, which give the appearance of citizen action groups but were apparently set up by Bivings and others on behalf of corporations. After leaving Monsanto, Byrne was quoted as saying, "think of the internet as a weapon on the table. Either you pick it up or your competitor does, but somebody is going to get killed."[54]

Chapela survived the attempts at character assassination, but is still feeling the effects. The university committee that has delayed for more than a year the decision to grant him tenure, receives letters from all over the world trying to convince them not to keep him at UC Berkeley. Chapela, like Pusztai before him, has been made a public example of what can happen to a scientist who breaks ranks with the pro-biotech mainstream.

Chapela says, "It's very hard for us to publish in this field. People are scared." Although pro-biotech scientists challenged his assertion that the CaMV promoter may create hotspots and scatter around the DNA, Chapela says that they are afraid to do the tests to support their contention. He says, "Have you wondered why people haven't come out to challenge our results?" Chapela says there is a de facto ban on scientists "asking certain questions and finding certain results. Who is being anti-scientific?"

On April 18, 2002, just two weeks after *Nature* partially rescinded Chapela's article, the Mexican government announced that massive genetic pollution of traditional maize varieties had indeed occurred in both the states of Oaxaca and next door in Puebla. Jorge Soberon, executive secretary of Mexico's biodiversity commission, admitted that the level of contamination "was far worse than initially reported".[55] They found GM DNA in up to 95 percent of maize plots tested. On average, 10 to 15 percent of the plants had GM kernels, with one field testing at 35 percent contamination. Genetic pollution had occurred, according to Soberon, "at a speed never before predicted. This is the world's worst case of contamination by genetically modified material because it happened in the place of origin of a major crop. It is confirmed. There is no doubt about it."[56] This news made headlines in Europe and Mexico. According to *Biodemocracy News*, in the US and Canada it was virtually ignored by the media.[57]

Changed Mice

Dutch undergraduate Hinze Hogendoorn, from University College, Utrecht, offered his mice a choice between GM and non-GM maize and soya altogether. Over a nine-week period, the mice consumed 61 percent non-GM and 39 percent GM food.

Hogendoorn then changed his experiment, to look for differences between a group fed GM food and another fed natural food. The GM-fed group ate more, probably because they were slightly heavier on average to begin with, but curiously, they gained less weight. In fact, by the end of the short experiment, they actually lost weight. On the other hand, the mice fed the non-GM diet ate less but gained more weight, continuing to gain weight until the end of the experiment. The results were statistically significant. In addition, one of the mice in the GM cage was found dead at the end of the experiment.[1]

The weight loss effect has also been observed elsewhere. Writer Steve Sprinkel, for example, had been told "about cattleman who saw the weight-gain of his cattle fall off when he switched over to GMO sources."[2] Tom Wiley of North Dakota described another difference: "I saw an advertisement from a farmer who was looking for non-Bt corn, as he was getting lower milk yields from the cattle that were eating Bt corn."[3]

Chapter 8

Changing Your Diet

In 1996 Greg Bretthauer was offered what he thought would be a great job—the dean of students at the Central Alternative High School in Appleton, Wisconsin. But when he visited the school and saw what it was like, he didn't want anything to do with it. Teens were "rude, obnoxious and ill-mannered"[1] he reported, and the school was out of control. They needed a police officer on staff to deal with discipline and weapons violations.

But in 1997, the school began to change almost miraculously, thanks to Barbara Reed Stitt, author of *Food and Behaviour, A Natural Connection*. She had learned about the profound effects of food in the unlikely position of probation officer. The first thing she did with anyone who came under her care was to change their diet. Time and time again their lives turned around.

"Over 80 percent of the probationers who appeared before me from 1970 through 1982," says Stitt, became "healthy, productive" members of society "after I started the dietary therapy programme." The changes in their lives were so apparent, one judge would regularly say to new probationers, "I'm going to send you down to Barbara Reed, and you're going to stay on the diet she gives you. If you don't, you'll be back in trouble—and next time you're going to jail!"[2]

Stitt was convinced that many of the problems facing schools in America are due to poor diets. Having changed the behaviour of criminals, she believed that influencing high school kids would be a piece of cake—or lack of it.

She and her husband Paul, a biochemist, approached their local school with an offer that was as unusual as it was generous. Take out the vending machines, take out the processed foods and feed the students fresh, whole, nutritious food and watch their behaviour improve. And the Stitts will pay the bill. In fact, since the Stitts owned Natural Ovens, a whole foods bakery, their company would send the school plenty of its own healthy fare AND place one of their own cooks on site at the school's kitchen.

School officials readily accepted the no-strings-attached offer and even expected to see a few changes. What they got was a revolution.

The school is calm, the kids are well behaved, truancy isn't an issue, and arguments are rare. Grades have improved and teachers are able to spend their time teaching instead of constantly disciplining. "I don't have the disruptions in class or the difficulties with student behaviour that I experienced before we started the food programme,"[1] says teacher Mary Bruyette.

Even students who were clearly headed for trouble have been swept up in the revolution. The numbers are impressive. Since the programme started five years ago, there have been no incidents of weapons, dropouts, expulsions, suicides, or even drugs.

Bretthauer visited the school four years after he had turned down the dean of students position and was amazed at what he saw. He says, "I happened to come back to interview for a different job and found the atmosphere entirely different."[3] He decided to take the job as dean of students after all.

Other schools in the district are asking for similar changes. Einstein Middle School made some modest changes and right away the kids were "more alert and focused", according to the principle. A middle school science teacher reported, "I've taught here almost thirty years. I see the kids this year as calmer, easier to talk to. They just seem more rational. I had thought about retiring this year and basically I've decided to teach another year—I'm having too much fun!"[1]

Students notice the changes that come with a healthier diet. One girl said, "Now that I concentrate I think it is easier to get along with people 'cause now I'm paying attention to what they have to say and not just worrying about what I have to say to them." Another student said, "If you're going for a big test you want to eat great."

"They have learned that with healthier foods it's going to make them a better person," says superintendent Thomas Scullen. "It keeps them more focused and makes them happier."[1] Many students have become advocates of healthy foods.

The news of the school's transformation has resulted in inquiries from around the world. They get requests daily on their website and have been featured on "Good Morning America", New Zealand radio and an Italian magazine, to name a few.

A teacher in the school district conducted a similar experiment with mice years earlier. In one cage, three mice ate junk food; in the other, three mice ate whole foods. The difference between the two groups was shocking.

The junk food mice, according to Stitt, "became very solitary and unsociable."[4] When they did interact, they would often fight.

Each cage had a cardboard tube taken from the inside of a paper towel roll. The junk food mice ripped theirs to shreds while the whole food group curled up inside their tube to sleep. The junk food mice also seemed to abandon their normal nocturnal behaviour, running around so much during the day that the teacher had to put a cover on the cage to keep the noise down in her class.

After two months on the junk food, two of the mice killed and ate the third.

At the end of the three-month experiment, the two remaining junk food mice were fed whole foods. In about three weeks, their behaviour once again became calm and gentle.

When teacher Sister Luigi Frigo heard about the mice experiment, she decided to repeat it with her second grade class in Cudahy, Wisconsin, and has done so every year for seven years. To protect the mice, however, she limits the junk food diet to only four days. First, she and her children observe the mice for one week, writing notes on their behaviour. Then three of the mice are fed items such as sugar-coated cereal, candy or cookies and diet soda. By the next day, she says, "Their behaviour changes drastically."[5] The junk food mice change from social, highly active animals to lazy, antisocial ones. They groom themselves more, appear nervous, hide their food, and can no longer perform some of the "tricks" they did beforehand. It takes about two to three weeks on the whole food diet for them to recover from the junk food. Once, the class tried to repeat the experiment on the same mice months later, but the animals refused to eat the junk food.

Mind-Altering Food

Although Stitt was not focused on removing genetically engineered foods per se, by taking out the vending machines, preparing most foods from scratch, and using olive oil instead of vegetable oils, her programme eliminated almost all the GM sources on the menu.

It is unclear which foods had contributed to behaviour problems. What is clear is that food can have a profound effect on behaviour, mood, happiness, and our entire quality of life. In 2002, research demonstrated that "food molecules act like hormones, regulating body functioning and triggering cell division. The molecules can cause mental imbalances ranging from attention-deficit and hyperactivity disorder to serious mental illness."[6] Food can be "more powerful than drugs". The researchers also said that food could alter "genes that affect whether we get cancer, heart disease, depression, schizophrenia or dyslexia." Eating the right foods not only extends our lives, they said, but "more importantly, we would maintain a higher quality of life as we age."

A study by the UK's Asthma and Allergy Research Centre supports this conclusion. For two weeks, 277 three-year-olds were fed fruit juice dosed with a total of 20 mg of four artificial colours along with the preservative sodium benzoate. These amounts are well below levels allowed in children's food and drink. For another two-week period, the children received plain fruit juice.

The parents were not told which two-week period their kids got the plain juice and which period they got the spiked juice. They were asked to write reports on their three-year-olds' behaviour. They covered areas such as "interrupting, concentration, disturbing others, difficulty settling down to sleep, fiddling with objects and temper tantrums." The data showed significant differences between the two trial periods. In fact, the study concluded that food colouring is the likely cause for one in four temper tantrums among the general child population.

Researchers said, "Significant changes in children's hyperactive behaviour could be produced by the removal of food colourings and additives from their diet." The benefits, they said, "would accrue for all children from such a change, and not just for those already showing hyperactive behaviour or who are at risk of allergic reactions."[7]

If additives and junk food have such a powerful affect on infants, students and probationers, how much of our irritability, distractions, restlessness, insomnia, anger, or depression are being dictated by what we eat? Science does not yet have the answer. Food's impact on mental and emotional health is not evaluated in traditional food safety assessments. And no research has yet looked at these effects related to GM foods.

One experiment, however, did come up with something quite by accident. A Dutch student who fed one group of mice GM maize and the other group natural maize discovered more than just a weight difference between the two. There were marked behavioural disparities as well. The mice fed GM food "seemed less active while in their cages". And when the mice were weighed at the end of the experiment, the GM-fed mice were "more distressed" than the other mice. According to the researcher, "Many were running round and round the basket, scrabbling desperately in the sawdust and even frantically jumping up the sides, something I'd never seen before."[8]

Certainly this single observation is an insufficient basis on which to draw conclusions about the effects of GM foods on the human psyche. On the other hand, any conclusions that there are *no* psychological effects are similarly unfounded.

GM Foods currently on the market

The studies and experiences described above provide compelling reasons to choose our food wisely. We may have had the objection: "Since we're all going to die anyway, why worry about my diet just to live a little longer?" But dietary changes can dramatically improve the quality of our lives, irrespective of lifespan or even health.

For those embarking on removing GMOs from their diet, the good news is that in the same stroke you can remove mood-altering artificial additives. That's because processed foods often contain both.

Currently, the major genetically engineered crops are soya, cotton, oilseed rape (canola) and maize. Other modified crops include some US zucchini (courgette) and yellow squash, Hawaiian papaya, and some tobacco. There may also be some remaining GM potatoes in the form of starch, but Monsanto is no longer marketing them. The GM tomatoes have similarly been taken off the market. (There are some reports that GM tomatoes may have been commercialized on a small scale in China.) US dairy products may contain milk from cows injected with rbGH, but this is banned in Europe. Even honey and bee pollen can contain GM sources.

In addition, there are genetically modified food additives, enzymes, flavourings and processing agents in thousands of foods on the grocery shelves, as well as health supplements. For example, the rennet used to make cheese is often a genetically engineered version. Aspartame, the diet sweetener, is a product of genetic engineering. And GM bacteria and fungi are used in the production of enzymes, vitamins and processing aids.

How to Avoid GM Foods in the UK

Avoiding GM foods is considerably easier in the UK than in the US. Only food ingredients from varieties of genetically modified soya, maize and oilseed rape have been approved for food use in Europe—and very little of these are actually used. GM tomatoes were sold as tomato purée in Safeway and Sainsbury stores between 1996 and 1999, when they were withdrawn.

In July 2003 the European Parliament passed laws requiring that food, derivatives (such as corn syrup and oil), and animal feed containing GM materials must be labelled "This product is produced from GMOs". Meat or dairy products from animals fed GM feed do not have to be labelled. These come into force on April 18, 2004.

Any intentional use of GM ingredients at any level has to be labelled. However, if small amounts (below 0.9% for EU-approved GM varieties, and below 0.5% for currently EU-unapproved but scientifically assessed varieties) of GM ingredients are accidentally present in non-GM ingredients, they do not have to be labelled. The European Parliament is tightening up legislation on GM foods: in 2003 the amount permitted in food that need not be labelled as containing GM ingredients was reduced from 1% to 0.9%. The approval for any new GM variety to be introduced into the market has to be agreed by all EU member states.

However, the British government continues to argue for less stringent laws on labelling, and is trying hard to overcome the widespread public antipathy to GM. The "10 Downing Street" website says that "New technologies play an increasing role in food production, and genetically modified foods are at the forefront of the changing nature of our food culture."

The best general strategy to avoid GM is to buy certified organic foods wherever possible, or wholefood products that are guaranteed GM-free. Processed foods are more likely to include ingredients that have been genetically modified.

Soya and Maize (corn) Derivatives
Most processed foods contain soya and/or maize in some form: see the list in Appendix A. In Europe, most of these (other than non-food ingredients) will be labelled if they consist of, contain or are produced from GMOs, whereas in many countries (such as the US) they won't be.

Vegetable Oils
Most generic vegetable oils and margarines used in restaurants and in processed foods are made from soya, maize, oilseed rape (canola), palm oil, sunflower or cottonseed. Significant portions of each of maize, soya, oilseed and cotton crops are genetically engineered and usually mixed together with

their non-GM counterparts before being pressed into oil. Different vegetable oils may be used in the same product, depending on price and availability.

For people who can't find non-GM vegetable oils, substitute oils include olive, sunflower, safflower, butter (see dairy below), almond and just about any other oil available. At present oilseed rape from the EU should produce GM-free oil as no GM oilseed is licensed for commercial growing.

Fruit and Vegetables
Many brands of dried fruit (including those in breakfast cereals) such as dates, currants, raisins and sultanas, are coated with oil from GM Soya, and as such will soon be labelled in the EU.

Bread and Bakery Products
When buying bakery products, avoid those that include "flour improver" and "flour treatment agent", as these may include GM-enzymes and additives. Some strains of the enzyme Alpha Amylase, which may be listed on bakery product labels, are genetically engineered.

Dairy Products
Dairy products from animals fed GM soya and maize will not be labelled as such. Organic dairy produce does not permit GM animal feed.

Honey
Many of the cheaper brands of honey sold in the UK are the produce of several countries, including Canada. Some Canadian honey comes from bees collecting nectar from canola (oilseed rape). Where the label says that honey is "imported honey" or "product of more than one country", it may include honey from Canada that originates from canola (around 60% of Canadian-grown canola is GM). It has been reported that this may be filtered of pollen to remove detectable GM DNA, then mixed with honey from other countries; it is still derived from GM plants.[9] In the EU, honey will not require labelling as GM, even if bees are feed on GM pollen. Organic UK honey is not currently on the market, but imported organic honey is available.

Meat and eggs
Organic meat and eggs come from animals that have been raised without hormones and with feed that is non-GMO. Meat from animals fed GM soya and maize will not be labelled as such.

Imported foods

Imported foods from America and Canada are generally suspect, because these countries have already embraced GM technologies. GM products include ice-cream, milk, milk powder, butter, soy sauce, chocolate, popcorn, chewing gum, health foods and vitamins. In the UK no GM fruits or vegetables are currently on sale, although the approval for GM tomatoes in 1996 (see above) has never been withdrawn.

Rapeseed Oil and Mutagenesis

Canola (rapeseed) oil has an interesting history. It was derived from rapeseed, which is normally toxic. Scientists changed the rapeseed through mutagenesis. This is done by subjecting the plant to radiation in order to promote mutations of the DNA. (This is not the same process known as food irradiation, which is used for killing microorganisms.)

After exposing the rapeseed to radiation, scientists studied the resultant mutated varieties and identified one that produced less of the dangerous toxin normally found in the plant. This new variety is called canola-named for Canada, where the variety is primarily grown.

Mutagenesis does not involve inserting genes into the DNA. Advocates insist that the radiation simply accelerates the normal process of mutation that occurs as part of a species' evolution and natural selection process. Others are suspicious of mutagenesis in general and of canola in particular. They avoid rapeseed oil because of its mutagenic origin and what they consider to be insufficient support for its claims of safety.

The focus of this book, however, is on GM foods that use gene insertion. We won't enter into the details of the canola/rapeseed controversy. For your information, if a brand of canola/rapeseed oil says non-GMO or organic, it means that there is no foreign gene inserted into the plants' DNA. However, the canola/rapeseed still has a mutagenic origin.

GM Additives, Cooking Aids, Vitamins and Enzymes

Genetic engineering is used in the production of many food additives, flavourings, vitamins, and processing aids, such as enzymes. According to the *Non-GMO Source*, "Such ingredients are used to improve the colour, flavour, texture and aroma of foods and to preserve, stabilize and add nutrients to processed foods. The number of minor ingredients that may be derived from GM sources, such as maize or soya, or produced using GMOs is vast."[10]

Among vitamins, vitamin C (ascorbic acid) is often made from maize, vitamin E is usually made from soya. Vitamins A, B2, B6 and B12 may be derived from GMOs as well. In addition, vitamin D and vitamin K may have "carriers" derived from GM maize sources, such as starch, glucose and maltodextrin. In addition to finding these vitamins in supplements, they are sometimes used to fortify foods. Organic foods, even if fortified with vitamins, are not allowed to use ingredients derived from GMOs.

Flavourings can also come from maize or other GM sources. For example, "Hydrolyzed vegetable protein (HVP), a commonly used flavour enhancer derived from corn [maize] and soy could be GMO,"[10] says the *Non-GMO Source*. Vanillin can also be GM.

The ingredients in some health food supplements, vitamins and medicines may be produced by biotechnology, so check with the manufacturer. Riboflavin (Vitamin B2) produced from GM-organisms has been approved for use in the UK. Riboflavin is used in baby foods, breakfast cereals, soft drinks, slimming foods etc. Under new legislation due in April 2004, food aids will not need to be labelled.

Many brands of dairy products, cereals, jam, fruit juice, cooking oil, sweeteners, slimming foods and beverages (including wine and beer) are produced with GM-enzymes. If you have any doubts about a particular product, ask the manufacturer.

"Genetically engineered bacteria and fungi are routinely used as sources of enzymes for the manufacture of a wide variety of processed foods,"[11] says GEO-PIE. The live organisms are not added to the foods themselves. Rather, they are grown in vats and produce large quantities of enzymes. The enzymes are removed, purified, and used in food production. The enzymes often get destroyed during the cooking process and are not present in the final product. As such, they are rarely listed on the label. Genetically altering bacteria and fungi has been going on since the 1980s. L-tryptophan is an example of how natural bacteria used in food or supplement production were modified to generate more product at less cost. One common enzyme is called chymosin, which is used in the production of hard cheeses. In the past, it was taken from the stomach linings of calves (called rennet). It is not allowed in organic cheese. Xanthan gum is another product that may be derived from a GM process.

According to the *Non-GMO Source*, "Europe, which is known for sensitivity about GMOs, is more advanced in genetic engineering of enzymes and microorganisms. The newest GM food labelling laws proposed by the

European Parliament call for labelling of food additives and colourings, but not processing aids such as enzymes."[10] This would also be true for yeasts.

GM yeasts have also been approved, but are not currently being used for food production. Avoiding GM additives is difficult, since the label will rarely list them. They are used in a wide variety of products including beer, alcohol, starch, dextrose, high fructose corn syrup, fruit juice, baked goods, sugar, malt syrups, bread, diet sweetener (aspartame), mayonnaise, cheese and other fermented dairy products, oils, fats and animal feed. The easiest ways to avoid these are to either buy products listed as organic or non-GMO, or prepare your own foods from basic, unprocessed ingredients. There's a list of current genetically modified enzymes in appendix B, describing how each is used. As we learn which food brands are made from GM bacteria and fungus, we will list them at www.seedsofdeception.com.

What Does a Non-GMO Label Mean?

There are currently no European regulations governing the use of the label "GM-free" or "Non-GMO". Each manufacturer can establish its own criteria. Some use the label if there are no ingredients from crops that have been genetically altered: no soya, maize, and so on. These same foods might contain GM processing agents, but probably do not.

Some foods with soya or maize products are labelled "Non-GMO" because the crops were grown from non-GM seed. But non-GMO seeds and crops can be contaminated. Therefore, each manufacturer decides how much vigilance is used to support that claim. Some companies rely only on affidavits by farmers. Others test their products.

One common test is an on-site "strip test". Like a home pregnancy test, a strip is dipped into a test tube containing a special solution mixed with the powdered crop. It will change colour if there are GMOs present. These strip tests are not effective for processed foods and have an inconsistent track record. While they claim to be sensitive enough to detect as little as 0.1 percent GM content, a published study that evaluated strip tests performed by grain handlers found that they missed detecting soybeans with 1 percent GM content about one-third of the time.

They are generally effective at detecting large percentages of GMOs. The more rigorous non-GMO claims are based on a test known as polymerase chain reaction or PCR. When used by skilled technicians, PCR can accurately detect GM content as low as 0.01 percent, and can determine what percentage of GMOs are present.

Irrespective of the test methods, manufacturers must select a level of GMO contamination that they consider acceptable. Having a zero tolerance is neither practical nor possible to guarantee.

Does Organic Mean Non-GM?

Yes. Organic standards ban genetically modified organisms (GMOs) and their derivatives. This prohibition extends to all areas of organic farming, including animal feed and veterinary products, and all areas of food processing, including additives and processing aids.

In the UK, organic food is defined as:

- having been produced to organic standards, which restrict the use of artificial pesticides and fertilizers
- from animals reared without the routine use of antibiotics and growth promoters
- Non-GM

EU regulations on organic food require that anyone who wishes to produce organic food must register with a certification body. In the UK, there are fifteen such bodies. The biggest is the Soil Association, which certifies over 70 per cent of the organic food sold in the UK. By law, a product that says 'organic' on the label (in the UK and the European Union) must contain 95% organic ingredients by weight. The other 5% can come only from a special list of tested non-organic ingredients, none of which contain GM ingredients.

Big Plans—Genetically Engineering the Food Supply

There are hundreds of GM products in the pipeline awaiting further development, approval, or commercialization. Virtually every type of popular produce has been genetically engineered in the lab. Some of the these include: wheat, rice, melons, cucumbers, strawberries, broccoli, grapes, sunflower, sugarcane, sugar beet, apples, lettuce, radicchio, carrots, coffee, cranberries, eggplant, oats, onions, peas, pineapples, plums, raspberries, sweet potatoes, walnuts and barley.

Many varieties are undergoing field trials in one of the thousands of test plots being conducted yearly.

Whatever risk any particular GM product has, great or small, the act of introducing so many varieties in the future can multiply a small risk into a virtual certainty.

Eating out

In Europe, restaurants are supposed to list foods which contain GM ingredients, or have the information available upon request. One eating establishment, for example, offered the following policy statement in 1999:

> "In response to concern raised by our customers . . . we have decided to remove, as far as possible, genetically modified soy and maize (corn) from all food products served in our restaurant. We will continue to work with our suppliers to replace GM soy and maize with non-GM ingredients. . . . We have taken the above steps to ensure that you, the customer, can feel confident in the food we serve."[12]

The statement was in reference to the cafeteria of Monsanto's UK headquarters in High Wycombe.

A Strange Deal

Responding to statements by US Trade Representative Robert Zoellick attacking the EU's stance on GM foods, European Development Commissioner Poul Nielson said, "This is a strange discussion. Very strange. We are approaching a point where I would be tempted to say I would be proposing a deal to the Americans which would create a more normal situation.

"The deal would be this: If the Americans would stop lying about us, we would stop telling the truth about them."[1]

Chapter 9

What You Can Do

I was on a lunch break from a swing dance workshop in St. Louis when some other dancers entered the restaurant. We asked them to join us. Over lunch I asked the man across from me what he did besides dancing. He said he worked for Monsanto as a molecular biologist, genetically engineering food.

I chewed slowly and considered my options.

We were, after all, eating lunch. And he was, after all, a fellow swing dancer. I decided to be gentle.

After some light friendly chatter about the potential allergenicity of genetically engineered constructs, I posed the question: When you blast the gene into the DNA, how do you know if you have disrupted a sequence that is important?

He said that they knew the sequence of many genes and tried to avoid inserting new genes in the middle of an existing one. After a pause, he added, we're continuing to learn which sequences in the DNA are important.

I asked: What if the whole DNA sequence is important? The theory behind genetic engineering assumes that the DNA is a bunch of discrete genes, all working independently, which, when added together, produce a plant, animal, or human. But that's not how the rest of the body functions and it's not how the ecosystem works either. They both involve complex, interrelated systems that we do not fully understand. Dangerous side effects of medicine and environmental disasters are often the result when we ignore that complexity and attempt to make a single change in isolation. That's where we get into trouble.

What if the sequence of the DNA operates in a holistic manner, perhaps using laws of nature we haven't yet discovered? Suppose, for example, the double helix structure of the DNA takes advantage of subtle laws of quantum mechanics or field effects. Couldn't disrupting a portion of the DNA sequence have unforeseen consequences that we have no idea even how to test for?

He was silent. The table was silent. We all just looked at our food for a while and then continued eating. Someone said, "That was deep." More silence.

The biologist then responded, "But you do know we need genetic engineering?"

"What?"

"You know we need genetic engineering."

"How so?" I asked.

"To feed the world."

And then he proceeded to give me the numbers. By the year two thousand and such and such, the world's population would be such and such. And there's no way that we can feed our planet with the current system of agriculture. . . .

As the Monsanto scientist was speaking, I knew he was sincere. I knew he believed deeply in what he was saying. And I knew he was wrong.

"Feeding the hungry" is described in the book *Dinner at the New Gene Café* as "the Final Argument".[1] After you effectively counter arguments that the technology is precise, the FDA has proven it safe, and it's just like traditional crossbreeding, in the end the discussion will come to the moral imperative that we need GM foods to feed the world.

Those who study the issue, however, say that the argument made by the biotech industry is based on propaganda, not on science. The organization Stop Hunger Now says, "Abundance, not scarcity, best describes the world's food supply." The truth is we have more food per person than any time in history—4.3 pounds per day.[2] A report from the United Nation's Food and Agriculture Organization (FAO) confirms that with food production increasing and population growth rate decreasing, we won't be running out.[3] The sad fact is that starvation is not generally associated with lack of food in the world. We have one and a half times the amount of food needed to feed the world, but people still go hungry. But that's another story.

The story here is that the Monsanto scientist sitting in front of me believes his industry's PR stance entirely—and with a quarter of a billion dollars being spent on getting this message across, so do countless others.

I learned more about how the industry drives their points home as I listened to their representatives address an agricultural biotechnology conference later that year. Each described another glorious breakthrough about genetic engineering and how it would solve the problems of agriculture. But whenever anyone brushed lightly on the topic of public resistance to GM foods, they all said the same sentence: "It's not a food safety issue." Each speaker would characterize the arguments against GM foods as cultural, or religious, or philosophical, or anti-science, or complicated, or a trade barrier, or anti-American. But, "of course, it's not a food safety issue."

Audience members were from agriculture, the food industry, academia and the media. I wondered how they were reacting to what was being said. During a break, I started up a conversation with a graduate student doing research on the sociological issues surrounding GM foods. As she shared some details of her work, she referred to the resistance to GM foods expressed in Europe and elsewhere. She quickly added, "Of course it's not a food safety issue."

Bingo. It had worked. The words she used, even the way she said them, mirrored precisely that of the previous speakers. They had another convert.

Getting Heard—Making a Difference

The graduate student believed that there is no food safety issue with GM foods and the Monsanto scientist at lunch believed GM foods could feed the starving. Monsanto's CEO, Robert Shapiro said, "Those of us in industry can take comfort. . . . After all, we're the technical experts. We know we're right. The 'antis' obviously don't understand the science, and are just as obviously pushing a hidden agenda—probably to destroy capitalism."[4]

If you challenge the industry, as Agriculture Secretary Glickman reported, "You're Luddites, you're stupid." Or worse, you are a scoundrel for turning a cold shoulder to the millions of starving in the world. Jack Kemp, former Republican nominee for vice president, had some choice words for those who called for safety testing and labelling of GM foods. He said they are, "ill-considered, anti-progress, left-wing, self-appointed . . . anti-technology activists."[5] It's not easy speaking against the pro-biotech current.

A revealing study demonstrated that a group of people trying to make a decision was swayed not by the suggestions of its most intelligent members, but by those who spoke the most. The biotech industry takes advantage of this principle in millions of dollars worth of ads and in years of pro-biotech media reports.

How, then, do we get a different message across? Major media has avoided covering the food safety issue. Even when GMO-related health issues are reported, the news is usually limited to short sound bites or a quoted opinion that is "balanced" by a pro-biotech quote challenging any concerns. General news stories are not sufficient. To convince someone that GM foods carry serious risks usually takes a prolonged discussion. It takes an even longer discussion to inspire someone to actually change his or her lifelong eating habits. That's where this book comes in. It's a portable long discussion—one that can be passed around. And it's unedited by the media and unsanitized by the industry.

Books have power. Upton Sinclair's novel *The Jungle* exposed the unsanitary conditions of the meat packing industry. After Teddy Roosevelt read the book on a long train trip, he pushed a bill through congress creating meat inspection. At a press conference, President Kennedy acknowledged the importance of Rachel Carson's book *Silent Spring*, which exposed the dangers of pesticides. Kennedy then had his scientific advisor look into the issue. The book was eventually "credited with beginning the American environmental movement, the creation of the Environmental Protection Agency, and the 1972 ban on DDT."[6]

Officials around the world who are in charge of GM food policy need to be made aware of the foods' dangers and of how their approval was based on politics, not science. They have been subjected to relentless promotion by the biotech industry and bullying by the US government to accept GM foods and crops. The revelations in this book might change that.

In the US, executives of large food companies may have a more immediate influence. This was exhibited in the UK in 1998, where the head of Iceland Frozen Foods sparked a revolution. After receiving several letters expressing concerns about GM foods, the company's chairman Malcolm Walker decided to find out what all the fuss was about. After learning about the issues, he ordered that GM soya and maize be removed from the company's house brand. Brochures denouncing GM foods were handed out at his chain of stores. Within half a year, the rest of the UK food industry followed suit. Executives from other chains acknowledged the influence of Iceland Frozen Foods on their decisions.

Iceland Frozen Foods later committed to sell only meat that had been fed non-GM animal feed. They also removed Monsanto's genetically engineered artificial sweetener aspartame from their products' ingredients. Their press release even referred to research linking aspartame to brain tumours.[7]

In the US, Whole Foods Market, Wild Oats and Trader Joe's announced that GMOs would be removed from their store brands. Gerber baby foods, as well as scores of health food products, have similarly changed their ingredients. (See www.seedsofdeception.com for a list.)

When a store or brand removes GM ingredients, it has a ripple effect through the industry. After a supermarket chain commits to eliminate GMOs, they usually send out a letter to their suppliers who in turn contact their suppliers and so on. A store may have hundreds of food items, each with a list of ingredients. Hundreds or thousands of businesses can be affected, right back to the farm level.

When a vendor receives a request to provide only non-GM ingredients, they usually test their products for GM content. If they make a change, they typically choose the minimum level of compliance necessary to meet the buyer's requirements. They'll remove only those GM ingredients specified, and establish the least costly testing and monitoring programme that their buyer will accept. Their choices are not motivated by food safety; it's economics—make a change or lose the customer. Buyers, therefore, are at the top of the food chain. They move the market. When McDonalds, Pringles and the other major potato buyers decided not to sell Monsanto's GM New Leaf potato, for example, it was soon taken off the market. McDonalds and others doomed Monsanto's potato because they wanted to satisfy consumer demands. We have that power.

European food chains likewise responded to consumer demands, and their switch to non-GMO products was a landslide. Once a few major manufacturers and chains announced their intention go non-GMO, no one wanted to be left out. This made it easier on the whole industry. All the vendors and ingredient suppliers switched to non-GM soya and maize at the same time.

The European food industry had to spend a lot of money to make the switch, often changing recipes to eliminate soya and maize altogether. They resented the biotech companies for the whole mess. After all, the food industry didn't ask for GM foods and did not benefit from them in any way. GM foods were not cheaper or more appealing. They were an expensive problem thrust on them from what they considered an insensitive and greedy American industry.

In an attempt to try to prevent a European-style revolt among the US food sector, the biotech industry has tried to align with the food industry, convincing food manufacturers that they are part of the family; an attack on biotech is an attack on food. The effectiveness of this strategy is illustrated by the Grocery Manufacturers Association's (GMA) ubiquitous presence in the media defending GM foods. It is no coincidence that Monsanto is a significant contributor to the GMA.

When StarLink hit in 2000, however, things got shaken up. Companies that spent millions in costly recalls began questioning their support for biotech and even publicly challenged loose government policies. Consumers were alerted to potential dangers and many Americans realized for the first time that they were eating GM foods.

In November 2002, the food industry scored another victory. Grains of maize that had been engineered to produce a vaccine for fighting a diarrhoea-causing virus in pigs was accidentally mixed into 500,000 bushels of soybeans in a Nebraska grain facility. The USDA ordered the soya to be destroyed and the maize's maker, Prodigene, to pay the $2.8 million bill. The fact that the contamination was even discovered was based on several coincidences and could easily have been missed. News reports of the incident also revealed that two months earlier, Prodigene had to destroy 155 acres of maize in Iowa, because wind-blown pollen from its drug-producing maize may have contaminated that as well.

Food companies realized that they had narrowly missed another StarLink. They were also alerted to the abysmal regulations governing the approximately 300 plots of GM plants engineered to produce medicine and industrial chemicals. Since 200 of these plots use maize, the possibilities of contamination due to cross-pollination and accidental mixing are great; many estimate the chances that some Americans have already eaten maize-grown pharmaceutical or chemical products at about 100 percent.

The US food industry is now clearly concerned. They realize how vulnerable they are to another StarLink-type recall, and they have some idea that the government is not adequately protecting consumers. The time may be perfect to create a US food industry landslide. Even one large company changing its policy could make GM foods unpopular very quickly. That is the thinking behind GE Food Alert, a coalition of seven organizations that have targeted America's largest food manufacturer, Kraft foods. Their campaign, described at www.krafty.org, is rallying consumers to contact Kraft, to ask the company to take out GM ingredients.

Please email or write food companies to share your concerns about GM foods. If you have stopped buying a food brand due to GMO issues, definitely let the company know. With your message, please suggest that they read this book; they'll learn about the health risks of GM foods and the significant liability they face by using them. You can download sample letters and emails at www.seedsofdeception.com.

For some manufacturers, switching to non-GMO ingredients is quite simple—there are easy-to-substitute alternatives. For others, substitutions are difficult. They may wait for the landslide.

Local Action

One of the easiest ways to effect such a landslide is to inspire change at the local level. Sometimes all it takes is a request. For example, I asked the owner of a local restaurant to take GM foods off his menu, explaining that there were several people in town that avoided them. He invited me into his kitchen to see what that would involve. He then switched from soya oil to olive and sunflower oils, replaced his zucchini (courgette) with an organic source and started using organic milk. Since his menu items used almost no packaged foods, the changes were simple and inexpensive. I wrote a short article about it for a local weekly paper, which he posted on his window. He saw an immediate increase in business.

Not to be outdone, a competing restaurant one block away also removed GM foods. I wrote an article for them as well. Two nearby restaurants then switched to non-GM oil and organic dairy. In fact, they raised the prices on a few entrées by 50 cents to cover the increased costs and posted signs explaining what they had done. Customers loved it. Now other restaurants in town are making the switch.

I never once had to discuss any safety issues about GM foods. It was enough for the restaurant owners to know that their customers preferred not to eat GM foods, or that a competitor was responding to that preference.

At www.seedsofdeception.com there are materials you can print out and give to restaurant owners that will explain the issue and help them to make a switch. There are even notes you can give to waiters or waitresses to help them accommodate your desire for non-GMO food.

Of all the local strategies, inspiring schools to make a change may be the most powerful. Schools throughout the UK and parts of Europe banned GM food years ago. In the 1990s, many Parent and Teacher Associations (PTAs) rallied against rbGH and more than a hundred US school districts banned milk from rbGH-treated cows. Wisconsin dairy farmer John Kinsman describes the method he used to inspire several schools. "I simply talked to parents of small children. Once mothers heard about this, they didn't rest until their school made the commitment."[8] Children are at greatest risk from the potential dangers of GM foods. Since there are few forces in nature stronger than a mother protecting her child, Kinsman's strategy is powerful. A Connecticut woman also found that having a member of the school board on her side was important. At www.seedsofdeception.com you will find sample letters you can use to make the approach to parents, board members and others easy and effective. Similarly, there are materials for college students to help them approach their campus food service.

According to information theory, "Knowledge has organizing power". Informed, knowledgeable people make a difference. In the USA, we know we can't depend on the our media or government to inform us. To get the stories in this book to the public, it has to be accomplished by its readers, passed from person to person. To make this affordable, I offer discounts for purchases of six or more copies of the American edition, and Green Books have agreed to offer a similar discount in the UK. Please buy as many as you can. See the order form at the end of this book.

Also, on my website you can sign up to become a member of the Institute for Responsible Technology. This is the organization I founded to help you stay up-to-date on the issues and to help make it easier to identify which foods are genetically modified. There is an electronic newsletter, information packets to download, and links to other organizations. We will also let you know when there are campaigns to contact elected officials to encourage support for important legislation.

What I Wrote, What I Didn't Write, Why

This book has focused on the issue of genetically modified food. It has not looked at gene therapy or genetically modified medicine. There are fundamental differences. Several scientists working in the field of gene therapy are appalled that genetic technology is being applied to food, which exposes our entire population and ecosystem to unnecessary risks. Gene therapy or GM medicine, on the other hand, may limit risks just to those individuals who agree in advance. I invite you to evaluate the other genetic technologies on a case-by-case basis.

This book also does not explore the most dangerous aspect of GM foods—the environmental impact. Once a genetically modified organism is released into the environment, it can never be recalled. Genes remain in the gene pool of a species, or pass between species for countless generations. The devastating environmental implications of GM foods are discussed in my forthcoming book, along with corresponding stories of government negligence and complicity.

The second book also describes how the biotech companies have taken advantage of farmers. The damage that this technology is doing to the farm sector is considerable—but well hidden. The revelations in the book will be a real shock. I will also introduce some new, important GM food safety issues. You can sign up at the website to be notified of the book's availability.

The narrow focus of this first book on the health risks of GM foods is intentional. It is designed to be a catalyst for change. I believe that for most people, legitimate concerns about food safety are far more effective at motivating changes in diet and buying habits than concerns for the environment, farmers, or other related issues. The narrow focus also makes it harder for any book reviewers and other book-related media to be diverted onto other topics. This *is* a food safety issue, and that needs to be aired in the press. As for the subjects of industry manipulation, incompetent science and government collusion, how else can one explain why these dangerous foods are on the market?

While I use the term "lies" to describe assertions that GM foods are safe, I do not believe that most people who make that claim are liars. Rather, I see them as buying into a myth. While it was spread intentionally, the myth has taken hold and is now a basic assumption of our food industry. It has also become the calling card of the US government.

Epilogue

When the US announced on May 13, 2003 that it would challenge the European Union's policy on GM foods through the World Trade Organization, US Trade Representative Robert Zoellick blamed fears of GM foods on "special interests that hype hysteria". He said, "millions of North Americans have been eating biotech food every day for years and not a single adverse health consequence has been documented." Zoellick said, "the EU bears a responsibility for ensuring that its health and environmental policies have a sound scientific basis." He added, "Sadly, as we've waited patiently for European leaders to step forward to deploy reason and science, the EU moratorium has sent a devastating signal to developing countries that stand to benefit most from innovative agricultural technologies." The following week, President Bush claimed that the EU has "blocked all new bio-crops because of unfounded, unscientific fears. . . . European governments should join—not hinder—the great cause of ending hunger in Africa."

Pro-biotech rhetoric is on the rise as the US attempts to force GM foods onto countries around the world. Major US media repeat the government's unsupported claims without question or analysis. In the UK, Tony Blair is similarly pushing the industry's agenda. On June 18, 2003 he said, "it is important for the whole debate to be conducted on the basis of scientific evidence, not on the basis of prejudice."

Michael Meacher, who had days earlier been replaced as the UK minister of the environment, responded to Blair with an article in the UK's *Independent*. Entitled "Are GM Crops Safe? Who Can Say? Not Blair," Meacher presented scientific evidence that countered the prime minister's vacuous safety assurances. The details of Meacher's argument, which have yet to be covered by a major US newspaper, highlight several of the points covered in this book.

Meacher wrote, "Contrary to the assurances of the biotech companies that genetic engineering is precise and simply extends traditional breeding techniques, it is actually quite different. When genetic crops are engineered, the gene is inserted randomly, out of a sequence that has evolved over hundreds of millions of years." Meacher explained that although engineers assumed that each gene creates a single protein, "the recent discovery that human beings have only some 30,000 genes . . . shows that this premise was wrong." Further, genes interact; one gene may trigger "other unpredicted and undesired effects". Meacher said, "The random position and lack of control of the gene's functions could change any character of the plant and might not be evident immediately." He cited examples of unexpected deformities in GM soya and cotton crops.

Meacher elaborated on the dangers of transferring allergies into a GM food, overuse of herbicides and the accidental switching on of a host organism's gene at random. He also discussed horizontal gene transfer—where genes can jump between organisms with unpredictable consequences. He said "the only human GM trial, commissioned ironically by the [UK's] Food Standards Agency" confirmed that GM DNA did, in fact, transfer to bacteria in the human gut. "Previously many scientists had denied that this was possible," he said. "But instead of this finding being regarded as a serious discovery which should be checked and re-checked, the spin was that this was nothing new and did not involve any health risk."

Meacher says that the while "it is often claimed that all GMOs have been 'rigorously tested,'" all that this testing amounts to is deciding whether a GM crop is similar in terms of its composition to the non-GM plant. . . . It wholly misses the point that health concerns are focused, not on known compounds, but on the effects of the GM technology which are unpredictable."

Meacher said, "The only Government-sponsored work ever carried on the health impacts of GMOs was Dr. Pusztai's work on rats and GM potatoes, and then, when it found negative effects, it was widely rubbished in government circles, even though his paper had been peer-reviewed six times before publication."

Meacher referred to a 2002 report by the Royal Society, which, although characteristically pro-biotech in its sentiment, did state that genetic modification "could lead to unpredicted harmful changes in the nutritional state of foods." They therefore recommended that potential health effects of GM foods be rigorously researched before being fed to pregnant or breast-feeding women, elderly people, those suffering from chronic disease, and babies. Meacher said, "Any baby food containing GM products could lead to a dramatic rise in allergies." Likewise, unexpected changes in oestrogen levels in GM soya used in infant formula "might affect sexual development in children", and that "even small nutritional changes could cause bowel obstruction." The article also cited a recent British Medical Association report that concluded, "there has not yet been a robust and thorough search into the potentially harmful effects of GM foodstuffs on human health."

"Finally," concludes Meacher, "it is often claimed by the biotech companies that there have been millions of people consuming GM foods over several years in the US, but without any ill-effects. However, there have actually been no epidemiological studies to support this claim. What is known is that coinciding with the introduction of GMOs in food in the US, food-derived illnesses are believed by the official US Center for Disease Control to have doubled over the past seven years. And there are many reports of a rise in allergies—indeed a 50 percent increase in soya allergies has been reported in the UK since imports of GM soya began. None of this of course proves the connection with GM, but it certainly suggests an urgent need for further investigation of this possible link. Typically, however, this has not been forthcoming."[1]

To Meacher's list of concerns, we can add the potential for cancer, toxins, new diseases and the other health effects discussed in relation to rbGH and the L-tryptophan disaster. There are also the numerous ways in which industry researchers apparently doctored their studies to avoid finding problems with GM foods. For example, Aventis heated StarLink maize four times longer than standard before testing for intact protein; Monsanto fed mature animals diets with only one tenth of their protein derived from GM soya; researchers injected cows with one forty-seventh the amount of rbGH before testing the level of hormone in the milk and pasteurized milk 120 times longer than normal to see if the hormone was destroyed; and Monsanto used stronger acid and more than 1,250 times the amount of a digestive enzyme recommended by international standards to prove how quickly their protein degraded. Cows that got sick were dropped from Monsanto's rbGH studies, while cows that got pregnant before treatment were counted as support that the drug didn't interfere with fertility; differences in composition between Roundup Ready soya and natural soya were omitted from a published paper; antibody reactions by rats fed rbGH were ignored by the FDA; deaths from rats fed the FlavrSavr tomato remain unexplained; and Aventis substituted protein derived by bacteria instead of testing protein taken from StarLink, among others.

A much longer list would be required to recount conflicts of interest, including job switches between government and industry, targeted campaign contributions, and the reliance by scientists, universities and research institutes on industry support.

One of the most dangerous aspects of the genetic engineering of food is the consistent attempt to silence those with contrary evidence or concerns. This book has introduced many who have been targets, including FDA scientists Richard Burroughs, Alexander Apostolou and Joseph Settepani; Health Canada scientists Shiv Chopra and Margaret Haydon; research scientists Arpad Pusztai, Ignacio Chapela and David Quist; authors Marc Lappé and Britt Bailey; physician Sam Epstein and the reporters who wrote about him; television reporters Steve Wilson and Jane Akre.

The attempts by these and others to alert the public and scientific community about the dangers of GM foods have yielded significant results. The rapid expansion of GM foods envisioned by Monsanto and others has slowed to a crawl, as more and more of the world community refuses to accept the foods or the rhetoric. On May 10, 2003, a new organization called the Independent Science Panel (ISP) was inaugurated at a conference in London. Committed to the "Promotion of Science for the Public Good", the founding members of ISP consist of twenty-four scientists in a wide variety of disciplines from all over the world. The ISP released a 136-page report entitled, "The Case for A GM-Free Sustainable World". This meticulously researched document summarizes the overwhelming evidence in favour of an immediate ban on GM foods. At the end of the document, the scientists recount the major problems characteristic of the GM food debate thus far.

As you read their list below, it may draw to mind examples that you have read in this book. I hope this will reinforce your confidence to voice these arguments yourself. Overturning a myth is not easy and cannot be accomplished by only a few individuals. Please join with those of us who are dedicated to getting the truth out, and doing what we can to protect our world and ourselves.

The scientists wrote: "We find the following aspects especially regrettable and unacceptable:

- Lack of critical public information on the science and technology of GM
- Lack of public accountability in the GM science community

- Lack of independent, disinterested scientific research into, and assessment of, the hazards of GM
- Partisan attitude of regulatory and other public information bodies, which appear more intent on spreading corporate propaganda than providing crucial information
- Pervasive commercial and political conflicts of interests in both research and development and regulation of GM
- Suppression and vilification of scientists who try to convey research information to the public that is deemed to harm the industry
- Persistent denial and dismissal of extensive scientific evidence on the hazards of GM to health and the environment by proponents of genetic modification and by supposedly disinterested advisory and regulatory bodies
- Continuing claims of GM benefits by the biotech corporations, and repetitions of these claims by the scientific establishment, in the face of extensive evidence that GM has failed both in the field and in the laboratory
- Reluctance to recognize that the corporate funding of academic research in GM is already in decline, and that the biotechnology multinationals (and their shareholders) as well as investment consultants are now questioning the wisdom of the 'GM enterprise'
- Attacks on, and summary dismissal of, extensive evidence pointing to the benefits of various sustainable agricultural approaches for health and the environment, as well as for food security and social well-being of farmers and their local communities."[2]

It's our food. Safe eating.

Appendix A
Genetically Modified Foods at a Glance

(Go to *www.seedsofdeception.com* for an up-to-date list.)

Plans to plant commercialized GM crops in the UK

At the time of publication of this book (Spring 2004), the commercial plant-ing of GM fodder maize has been given UK government approval subject to conditions being agreed to limit the amount of herbicide used on the crop. If this is agreed by the EU and EC, the decision to add the one GM fodder maize variety to the National List of Varieties (the Seed List) will have to be jointly agreed by the governments in Wales, Scotland and Westminster (act-ing for Northern Irelend). Without this approval, GM maize cannot be mar-keted in the UK. The government has decided against approving the other two GM crops (spring oilseed rape and sugar/fodder beet) evaluated in the recent trials.

Currently commercialized GM crops worldwide

In 2003, the number of countries responsible for 99 percent of the global biotech crop area expanded to six (up from four in 2002): the US, Argentina, Canada and China were joined by Brazil and South Africa. The main crops are soya, maize, oilseed rape (canola) and cotton.

Other sources of GMOs

Dairy products from cows injected with rbGH. Food additives, enzymes, flavourings, and processing agents, including the sweetener aspartame (NutraSweet®) and rennet used to make hard cheeses.
Meat, eggs, and dairy products from animals that have eaten GM feed.
Honey and bee pollen that may have GM sources of pollen.

Some of the ingredients that may be genetically modified

Vegetable oil (soya, maize, cottonseed, or canola [rapeseed]), margarines, soya flour, soya protein, soya lecithin, textured vegetable protein, cornmeal, corn syrup, dextrose, maltodextrin, fructose, citric acid, and lactic acid.

Some of the foods that may contain GM ingredients

Infant formula, salad dressing, bread, cereal, hamburgers and hot dogs, margarine, mayonnaise, cereals, crackers, biscuit, chocolate, sweets, fried food, chips, veggie burgers, meat substitutes, ice cream, frozen yogurt, tofu, tamari, soy sauce, soya cheese, tomato sauce, protein powder, baking powder, alcohol, vanilla, powdered sugar, peanut butter, enriched flour and pasta. Non-food items include cosmetics, soaps, detergents, shampoo, and bubble bath.

Appendix B
Enzymes Derived from GM Organisms

Reproduced with permission from GEO-PIE
http://www.geo-pie.cornell.edu//gmo.html

Enzyme Name	GM Organism	Use (examples)
alpha-acetolactate decarboxylase	bacteria	removes bitter substances from beer
alpha-amylase	bacteria	converts starch to simple sugars
catalase	fungi	reduces food deterioration, particularly egg-based products
chymosin	bacteria or fungi	clots milk protein to make cheese
cyclodextrin-glucosyl transferase	bacteria	starch/sugar modification
beta-glucanase	bacteria	improves beer filtration
glucose isomerase	bacteria	converts glucose sugar to fructose sugar
glucose oxidase	fungi	reduces food deterioration, particularly egg-based products
lipase	fungi	oil and fat modification
maltogenic amylase	bacteria	slows staling of breads
pectinesterase	fungi	improves fruit juice clarity
protease	bacteria	improves bread dough structure
pullulanase	bacteria	converts starch to simple sugars
xylanase (hemicellulase)	bacteria or fungi	enhances rising of bread dough

Notes

Introduction

1 Stuart Laidlaw, 'StarLink Fallout Could Cost Billions', *The Toronto Star*, January 9, 2001
2 Robert Cohen, *Milk, the Deadly Poison*, Argus Publishing, Englewood Cliffs, New Jersey, 1998, p. 133

Chapter 1

1 Unless otherwise indicated, quotes from Arpad Pusztai, his wife Susan, or Philip James are based on personal communications with Arpad Pusztai.
2 Project Censored, http://www.projectcensored.org/publications/2000/2001/7.html
3 *GM-FREE Magazine*, vol. 1, no. 3, August/September 1999
4 Transcript from 'World in Action' provided by Arpad Pusztai
5 John Vidal, 'Revolts Against Monsanto and Genetically Engineered Foods throughout Europe', *The Guardian* (UK), March 19, 1999, http://www.netlink.de/gen/Zeitung/1999/990220d.htm
6 Sheldon Rampton and John Stauber, *Trust Us We're Experts*, Jeremy P. Tarcher/Putnam, New York, 2001
7 Arpad Pusztai, http://www.freenetpages.co.uk/hp/A.Pusztai/memo.txt
8 Nigel Hawkes, 'Scientist's Potato Alert was False, Laboratory Admits', *The Times* (London), August 13, 1998
9 Euan McColm, 'Doctor's Monster Mistake', *Scottish Daily Record & Sunday Mail*, October 13, 1998, p. 6
10 'Peer review vindicates scientist let go for "improper" warning about genetically modified food', *Natural Science Journal*, March 11, 1999, http://naturalscience.com/ns/cover/cover8.html
11 Bill Lambrecht, *Dinner at the New Gene Café: How Genetic Engineering Is Changing What We Eat, How We Live, and the Global Politics of Food*, St. Martin's Press, New York, 2001, p. 230
12 Abi Berger, 'Hot potato', *Student BMJ April 1999: Medicine and the media*, http://www.studentbmj.com/back_issues/0499/data/0499mm1.htm
13 Bill Lambrecht, *Dinner at the New Gene Café*, p. 232
14 Ziauddin Sardar, 'Loss of Innocence: Genetically Modified Food', *New Statesman* (UK), vol. 129, no. 4425, February 26, 1999, p. 47
15 Testimony of Professor Philip James and Dr. Andrew Chesson, Examination of witnesses (Questions 207–219) March 8, 1999, http://www.parliament.the-stationery-office.co.uk/pa/cm199899/cmselect/cmsctech/286/9030815.htm
16 Sheldon Rampton and John Stauber, p. 172
17 Sarah Ryle, 'Food Furore: the Man With the Worst Job in Britain', *The Observer* (London), February 21, 1999
18 Fran Abrams, 'Parliament Food: "Cynical" Monsanto Branded Public Enemy Number One', *The Independent* (London), March 23, 1999

19 John Vidal and David Hencke, 'Genetic food in crisis', *The Guardian*, November 18, 1998
20 Gwynne Dyer, 'Frankenstein Foods', *Globe and Mail*, February 20, 1999
21 Marie Woolf, 'People distrust Government on GM foods', *Sunday Independent* (London), May 23, 1999
22 Friends of the Earth Press release, 'Supermarket Loyalty Cards to Track GM Food Threat', January 25, 1999, http://www.foe.co.uk/pubsinfo/infoteam/press-rel/ 1999/19990125154458.html
23 Geoffrey Lean, 'Exposed: Labour's real aim on GM food', *Sunday Independent* (London), May 23, 1999
24 Editorial, 'Less Spin, More Science', *Sunday Independent* (London), May 23, 1999
25 Laurie Flynn and Michael Sean Gillard, 'Pro-GM food scientist "threatened editor"' GM food: special report, *The Guardian*, November 1, 1999, http://www.guardian. co.uk/Print/0,3858,3923559,00.html
26 Editorial, *Lancet*, May 22, 1999, p. 1769
27 NLP Wessex, 'Survey of scientists and government ministers exposes complete lack of independent safety testing of GM foods; Independent safety tests of Genetically Modified foods have never been carried out', April 7, 2001, http://www.mindfully.org/GE/GE2/Survey-Scientists-Government.htm
28 Sheldon Rampton and John Stauber, p. 154
29 Ian F. Pryme and Rolf Lembcke, '*In Vivo* Studies on Possible Health Consequences of genetically modified food and Feed—with Particular Regard to Ingredients Consisting of Genetically Modified Plant Materials', *Nutrition and Health*, vol. 17, 2003, in press
30 Barbara Keeler and Marc Lappé, 'Some Food for FDA Regulation', *Los Angeles Times*, January 7, 2001
31 Arpad Pusztai, 'Genetically Modified Foods: Are They a Risk to Human/Animal Health?' June 2001, http://www.actionbioscience.org/biotech/pusztai.html
32 Joel Bleifuss, 'No Small (Genetic) Potatoes', *In These Times.com*, January 10, 2000
33 Noteborn and others, 'Safety assessment of the Bacillus thuringiensis insecticidal crystal protein (CRY1A9b) expressed in transgenic tomatoes', In: Engel and others (eds) ACS Symposium Series 605, 'Genetically Modified Foods—Safety Issues', American Chemical Society, Washington, D.C., pp. 135–147, 1995
34 Department of Animal and Poultry Sciences, University of Guelph, 'The Effect of Glufosinate Resistant Corn on Growth of Male Broiler Chickens', Report No. A56379, July 12, 1996
35 Greg Winter, 'Contaminated Food Makes Millions Ill Despite Advances', *New York Times*, March 18, 2001, http://www.btinternet.com/~nlpwessex/Documents/ cdcfood.htm
36 Mae-Wan Ho and Jonathan Matthews, 'Suppressing Dissent in Science with GM Foods', http://www.mercola.com/2001/mar/14/gm_foods.htm
37 Liz Lightfoot, 'Scientists "asked to fix results for backer"', *Times Higher Education Supplement*, Institute of Professionals, Managers and Specialists, September 8, 2000

38 Eyal Press and Jennifer Washburn, 'The Kept University', *The Atlantic Monthly*; March 2000; vol. 285, no. 3; pp. 39–54, http://www.theatlantic.com/issues/ 2000/03/press.htm

39 Ralph G. Walton, 'Survey of Aspartame Studies: Correlation of Outcome and Funding Sources', 1998, http://www.dorway.com/peerrev.htmll

40 Samuel Epstein and Pete Hardin, 'Confidential Monsanto Research Files Dispute Many bGH Safety Claims', *The Milkweed*, January 1990

41 Jaan Suurküla, M.D., 'Dysfunctional science: Towards a "pseudoscientfic world order?"' March 14, 2000, http://www.psrast.org/crisisofsci.htm

42 'World renowned scientist lost his job when he warned about GE foods', Physicians and Scientists for Responsible Application of Science and Technology (PSRAST), http://www.psrast.org/pusztai.htm

'Wisdom of the Geese'
1 Mark Newhall, 'He Says Geese Don't Like Roundup Ready Beans' *Farm Show*, vol. 24, no. 5, 2000

Chapter 2
1 Bill Lambrecht, *Dinner at the New Gene Café: How Genetic Engineering Is Changing What We Eat, How We Live, and the Global Politics of Food*, St. Martin's Press, New York, 2001

2 Barry Commoner, 'Unraveling the DNA Myth: The spurious foundation of genetic engineering', *Harper's*, February 2002, http://www.mindfully.org/GE/GE4/ DNA-Myth-CommonerFeb02.htm

3 A.S. Reddy and T.L. Thomas, 'Modification of plant lipid composition: Expression of a cyanobacterial D6-desaturase gene in transgenic plants', *Nature BioTechnology*, vol. 14, 1996, pp. 639–642

4 Michael Hansen, 'Possible Human Health Hazards of Genetically Engineered Bt Crops: Comments on the human health and product characterization sections of EPA's Bt Plant-Pesticides Biopesticides Registration Action Document', Presented to the EPA Science Advisory Panel Arlington, VA, October 20, 2000

5 T. Inose and K. Murata, 'Enhanced accumulation of toxic compound in yeast cells having high glycolytic activity: A case study on the safety of genetically engineered yeast'. International Journal of Food Science and Technology, vol. 30, 1995, pp. 141–146

6 'Making Crops Make More Starch', BBSRC Business, UK Biotechnology and Biological Sciences Research Council, January 1998, pp. 6–8

7 'Speaker Hastert Calls for End of European Union's "Protectionist, Discriminatory Trade Policies"', *US Newswire*, Washington, D.C., March 26, 2003

8 George Wald, 'The Case against Genetic Engineering' in The Recombinant DNA Debate, Jackson and Stich, eds. pp. 127–128 (Reprinted from *The Sciences*, Sept./ Oct. 1976)

9 Personal communication with Joseph Cummins

10 David Schubert, 'A different perspective on GM food', *Nature Biotechnology* vol. 20, 2002, p. 969

11 Mothers for Natural Law website, http://www.safe-food.org/-issue/scientists.html

12 Richard Strohman, professor emeritus, Department of Molecular and Cell Biology, University of California, Berkeley, 2000, http://www.mindfully.org/GE/Strohman-Safe-Food.htm

13 Gundula Meziani and Hugh Warwick, 'Seeds of Doubt', Soil Association (UK), September 17, 2002

14 Danny Penman, BBC Tomorrow's World Magazine, October 1998

15 Dale and others, 'Transgene expression and stability in Brassica', ACTA Horticulturae, 1998, no. 459, pp. 167–171

16 John Vidal, 'GM genes found in human gut', The Guardian, July 17, 2002, http://www.guardian.co.uk/gmdebate/Story/0,2763,756666,00.html

17 Ronnie Cummins and Ben Lilliston, 'Genetically Engineered Food: A Self-Defense Guide for Consumers', Marlowe and Company, New York, 2000

18 P. Meyer, F. Linn, I. Heidmann, H. Meyer, I. Niedenhof, and H. Saedler, 'Endongenous and environmental factors influence 35S promoter methylation of a maize A1 gene construct in transgenic petunia and its colour phenotype', Molecular Genes and Genetics, vol. 231, no.3, 1992, pp. 345–352

19 Mae-Wan Ho, Angela Ryan and Joe Cummins, 'Cauliflower Mosaic Viral Promoter—A Recipe for Disaster?' Institute of Science in Society, http://www.i-sis.org.uk/camvrecdis.php

20 Rob Edwards, 'GM expert warns of cancer risk from crops', The Sunday Herald, December 8, 2002, http://www.sundayherald.com/29821

21 Mae-Wan Ho and others, 'CaMV 35S promoter fragmentation hotspot confirmed, and it is active in animals', Microbial Ecology in Health and Disease, v. 13, 2000

22 Susan Benson, Mark Arax and Rachel Burstein, 'Growing Concern: As biotech crops come to market, neither scientists—who take industry money—nor federal regulators are adequately protecting consumers and farmers', Mother Jones, January/February 1997, http://www.motherjones.com/mother_jones/JF97/biotech_jump2.html

23 George Gallepp 'Scientists Find Compound that Makes Bt Pesticide More Effective', College of Agriculture and Life Sciences, University of Wisconsin, May 21, 2001, http://www.cals.wisc.edu/media/news/05_01/zwitter_Bt.html

24 ACNFP Review (Application to the UK Advisory Committee on Novel Foods and Processes), page 59, table 7, re: '3.5% corrected milk'

25 Marie Woolf, 'GM foods—Revealed: false data misled farmers', Sunday Independent, February 21, 1999

26 'US biotech researchers careless with 386 pigs—FDA', Reuters, February, 6, 2003

27 'Monsanto GM seeds contain 'rogue' DNA', Scotland on Sunday, May 30, 2000, http://members.tripod.com/~ngin

28 P. Windels, I. Taverniers, A. Depicker, E. Van Bockstaele, and M. DeLoose, 'Characterisation of the Roundup Ready soybean insert', European Food Research and Technology, vol. 213, 2001, pp. 107-112

29 Andrew Pollack, 'Mystery DNA Is Discovered in Soybeans by Scientists', New York Times, August 16, 2001

30 Alex Kirby, 'Greenpeace worried by "mystery DNA"', *BBC News Online*, August 15, 2001, http://news.bbc.co.uk/hi/english/sci/tech/newsid_1492000/1492939.stm
31 A. J. C. de Visser and others, 'Crops of uncertain nature? Controversies and knowledge gaps concerning genetically modified crops: An inventory', *Plant Research International*, Report 12, August 2000
32 Mothers for Natural Law website, http://www.safe-food.org/-issue/scientists.html

'Wisdom of the Cows'
1 Personal communication with Howard Vlieger
2 Gundula Meziani and Hugh Warwick, '*Seeds of Doubt*', Soil Association (UK), September 17, 2002
3 Steve Sprinkel, 'When the Corn Hits the Fan', *Acres*, U.S.A., September 18, 1999

Chapter 3
1 Anne McIlroy, 'Pierre Blais thought it was his duty', *Globe and Mail* (Canada), November 18, 1998
2 James Baxter, *The Ottawa Citizen*, October 23, 1998, p. A1
3 Steve Wilson, 'Secret Canadian Govt. Study Reveals Serious Faults with bGH Research; FDA Approval Was Based on Faulty Conclusion?' October 7, 1998, http://www.foxbghsuit.com/jasw1007.htm
4 Craig Canine, 'Hear No Evil: In its determination to become a model corporate citizen, is the FDA ignoring potential dangers in the nation's food supply?' *Eating Well*, July/August 1991
5 Jeff Kamen, 'Formula for Disaster', *Penthouse*, March 1999
6 Robert Cohen, *Milk, the Deadly Poison*, Argus Publishing, Englewood Cliffs, New Jersey, 1998
7 Judith C. Juskevich and C. Greg Guyer, 'Bovine Growth Hormone: Human Food Safety Evaluation', *Science*, 1990, vol. 249, pp. 875-884
8 Personal communication with Peter Hardin
9 Shiv Chopra and others, rBST (Nutrilac) 'Gaps Analysis' Report by rBST Internal Review Team, Health Protection Branch, Health Canada, Ottawa, Canada, April 21, 1998
10 Personal communication with Shiv Chopra
11 Peter Montague, 'Milk Controversy Spills into Canada', *Rachel's Environment and Health News*, no. 621, October 22, 1998, http://www.rachel.org/bulletin/bulletin.cfm?issue_ID=1168
12 Samuel Epstein and Pete Hardin, 'Confidential Monsanto Research Files Dispute Many bGH Safety Claims', *The Milkweed*, January 1990
13 Pete Hardin, 'rbGH: Appropriate Studies Haven't Been Done', *The Milkweed*, July 2000
14 Robert Cohen testimony before FDA panel, December 2, 1999
15 Robert J. Collier and others, '[Untitled Letter to the Editor]', *Lancet*, vol. 344, September 17, 1994, p. 816

16 T. B. Mepham and others, 'Safety of milk from cows treated with bovine soma-totropin', *Lancet*, Vol. 344, November 19, 1994, pp. 1445-1446

17 William H. Daughaday and David M. Barbano, 'Bovine somatotropin supple-mentation of dairy cows: Is the milk safe?' *Journal of the American Medical Association*, vol. 264, no. 8, August 22, 1990, pp. 1003-1005

18 C. G. Prosser and others, 'Increased secretion of insulin-like growth factor-1 into milk of cows treated with recombinantly derived bovine growth hormone', *Journal of Dairy Science*, vol. 56, 1989, pp. 17-26

19 Peter Montague 'Milk, rbGH, and Cancer', *Rachel's Environment and Health News*, no. 593, April 9, 1998

20 Toshikiro Kimura and others, 'Gastrointestinal Absorption of Recombinant Human Insulin-Like Growth Factor-I in Rats', *The Journal of Pharmacology and Experimental Theraputics*, vol. 283, no. 2, November 1997, pp. 611-618

21 Robert P. Heaney and others, 'Dietary changes favorably affect bone remodel-ing in older adults.' *Journal of the American Dietetic Association*, vol. 99, no. 10, October 1999, pp. 1228-1233

22 M. Lippman, 'Growth factors, receptors and breast cancer', *Journal of National Institutes on Health Res.*, vol. 3, 1991, pp. 59-62

23 E. A. Musgrove, and others, 'Acute effects of growth factors on T-47D breast cancer cell cycle progression.' *European Journal of Cancer*, vol. 29A, no. 16, 1993, pp. 2273-2279

24 June M. Chan and others, 'Plasma Insulin-Like Growth Factor-1 [IGF-1] and Prostate Cancer Risk: A Prospective Study', *Science*, vol. 279, January 23, 1998, pp. 563-566

25 S. E. Hankinson, and others, 'Circulating concentrations of insulin-like growth factor 1 and risk of breast cancer', *Lancet*, vol. 351, no. 9113, 1998, pp. 1393-1396

26 H. Yu and others, 'Plasma levels of insulin-like growth factor-I and lung cancer risk: a case-control analysis', *Journal of the National Cancer Institute*, vol. 91, no. 2, January 20, 1999

27 R. Torrisi and others, 'Time course of fenretinide-induced modulation of circu-lating insulin-like growth factor (IGF)-i, IGF-II and IGFBP-3 in a bladder cancer chemoprevention trial', *International Journal of Cancer*, vol. 87, no. 4, August 2000, pp. 601-605

28 S. E. Dunn and others, 'Dietary restriction reduces insulin-like growth factor I levels, which modulates apoptosis, cell proliferation, and tumour progression in p53-deficient mice', *Cancer Research*, vol. 57, no. 4, 1997, pp. 667-672

29 'Milk, Pregnancy, Cancer May Be Tied', Reuters, September 10, 2002

30 'New Study Questions rbGH Safety', *The Capital Times*, December 20, 1998

31 John Robbins, *The Food Revolution: How Your Diet Can Help Save Your Life and Our World*, Conari Press, Berkeley, California, 2001

32 Michael Culbert, *Medical Armageddon*, C and C Communications, 1997

33 James P. Carter, *Racketeering in Medicine: The Suppression of Alternatives*, Hampton Roads Publishing Company, Inc., Norfolk, VA, 1992, http://www.banned-books.com/truth-seeker/1994archive/121_2/ts212c.html

34 *San Francisco Chronicle*, January 2, 1970

35 Karl Flecker, 'Outside Advice? Are the External Panels on bGH Following Conflict of Interest Policy and Guidelines?' Council of Canadians report, Dec. 3, 1998, http://www.canadians.org/display_document.htm?COC_token=024PS24&id=37&isdoc=1
36 'Ottawa bans bovine growth hormone', CBC, January 15, 1999
37 Richard Wolfson, 'Update on Health Canada Scientists', April 20, 2000, http://www.freedomtocare.org/page145.htm
38 W. A. Knoblauch, 'The Impact of BST on Dairy Farm Income and Survival', Cornell University, Ithaca, New York, June 1992, http://www.nal.usda.gov/bic/BST/ndd/THE_IMPACT_OF_BST_ON_DAIRY_FARM_INCOME.html
39 Loren Tauer, 'Impact of BST on Farm Profits', Presented at the 4th annual conference on the 'Economics of Agricultural Biotechnology', Ravello, Italy, August 2000, http://aem.cornell.edu/research/researchpdf/wp0009.pdf
40 APHIS Info Sheet, December 2002, http://www.aphis.usda.gov/vs/ceah/cahm/Dairy_Cattle/Dairy02/bst-orig.pdf
41 Matt Wickenheiser, 'Oakhurst sued by Monsanto over milk advertising', *Portland Press Herald*, July 8, 2003
42 Michael Taylor, 'Interim Guidance on the Voluntary labelling of Milk and Milk Products From Cows That Have Not Been Treated with Recombinant Bovine Somatotropin', FDA Docket No. 94D-0025, February 7, 1994, http://www.idfa.org/news/stories/2002/12/fdarbstpolicy.pdf

'Wisdom of the Cows and Pigs'
1 Personal communication with Bill Lashmett

Chapter 4
1 Personal communication with Betty Hoffing
2 Story on Harry Schulte, WCPO-TV 9, 11 PM News, Cincinnati, Ohio, February 26, 1998
3 Janet O'Brien, *National EMS Network Newsletter*, Fall 1997
4 Bruce Frendlich, *National EMS Network Newsletter*, Fall 1996
5 Sheldon Rampton and John Stauber, *Trust Us We're Experts*, Jeremy P. Tarcher/Putnam, New York, 2001
6 Barbara Deane, 'Anatomy of an Epidemic', *Reader's Digest*, April 1991
7 Douglas L. Archer, Deputy Director, Center for Food Safety and Applied Nutrition, FDA, Testimony before the Subcommittee on Human Resources and Intergovernmental Relations Committee on Government Operations House of Representatives, July 18, 1991
8 Phillip A. Hertzman and others, 'The Eosinophilia-Myalgia Syndrome: The Los Alamos Conference', *Journal of Rheumatology*, vol. 18, no. 6, 1991, pp. 867-873
9 William Crist, Investigative report on L-tryptophan, found at www.biointegrity.org
10 Laurie Garrett, 'Genetic engineering flaw blamed for toxic deaths', *Newsday*, August 14, 1990, p. C-1

11 P. Raphals, 'Does medical mystery threaten biotech?' *Science*, vol. 249, no. 619, 1990

12 E. A. Belongia and others, 'An investigation of the cause of the eosinophilia-myalgia syndrome associated with tryptophan use', *The New England Journal of Medicine,* August 9, 1990

13 Philip Raphals, 'EMS deaths: Is recombinant DNA technology involved?' *The Medical Post*, November 6, 1990

14 Personal communication with Gerald Gleich, M.D.

15 Leslie A. Swygert and others, 'Eosinophilia-Myalgia Syndrome: Results of National Surveillance', *Journal of the American Medical Association*, October 3, 1990, vol. 264, no. 13, pp. 1698-1703

16 Edwin M. Kilbourne and others, 'Tryptophan Produced by Showa Denko and Epidemic Eosinophilia-Myalgia Syndrome', *Journal of Rheumatology Supplement*, vol. 23, no. 46, October 1996, pp. 81-92

17 National Eosinophilia-Myalgia Syndrome Network, position statement, approved quote by Gerald J. Gleich, Mayo Clinic and Foundation, May 25, 2000

18 Frank Silvestri and John Massicot, 'EMS Lawsuits', *National EMS Network Newsletter*, vol. II, issue 2, June 2001, p. 6

19 John Robbins, *The Food Revolution: How Your Diet Can Help Save Your Life and Our World*, Conari Press, Berkeley, California, 2001, p. 335

20 Danny Penman, 'GE—Of the top 100 economies 51 are multinational companies, the rest are countries', *BBC Tomorrow's World* magazine, October, 1998

21 L. R. B. Mann, D. Straton, and W. E. Crist, 'The Thalidomide of Genetic Engineering, GE issue of *Soil & Health* (New Zealand), August 1999

'Wisdom of Squirrels, Elk, Deer, Raccoons, and Mice'

1 Personal communication with Howard Vlieger

2 Gundula Meziani and Hugh Warwick, '*Seeds of Doubt*', Soil Association (UK), September 17, 2002

3 Steve Sprinkel, 'When the Corn Hits the Fan', *Acres*, U.S.A., September 18, 1999

4 Hinze Hogendoorn, http://www.talk2000.nl/mice/talk-Extended.htm

Chapter 5

1 Kurt Eichenwald and others, 'Biotechnology Food: From the Lab to a Debacle', *New York Times*, January 25, 2001

2 Michael Grunwald, 'Monsanto Held Liable for PCB Dumping', *Washington Post*, February 23, 2002

3 Sheldon Rampton and John Stauber, *Trust Us We're Experts*, Jeremy P. Tarcher/Putnam, New York, 2001, p. 164

4 Bush Library, http://bushlibrary.tamu.edu/research/find/foia/1999-0129-F.html 1999-0129-F

5 'Statement of Policy: Foods Derived from New Plant Varieties', *Federal Register* vol. 57, no. 104 at 22991, May 29, 1992

6 Steve Druker, www.biointegrity.org

7 Linda Kahl to James Maryanski, about Federal Register document '*Statement of Policy: Foods from Genetically Modified Plants*', January 8, 1992, www.biointegrity.org

8 Louis J. Pribyl, 'Biotechnology Draft Document, 2/27/92', March 6, 1992, www.biointegrity.org

9 Edwin J. Mathews to the Toxicology Section of the Biotechnology Working Group, 'Analysis of the Major Plant Toxicants', October 28, 1991, www.biointegrity.org

10 Samuel I. Shibko to James Maryanski, 'Revision of Toxicology Section of the *Statement of Policy: Foods Derived from Genetically Modified Plants*', January 31, 1992, www.biointegrity.org

11 Division of Food Chemistry and Technology and Division of Contaminants Chemistry, 'Points to Consider for Safety Evaluation of Genetically Modified Foods: Supplemental Information', November 1, 1991, www.biointegrity.org

12 Gerald B. Guest to James Maryanski, 'Regulation of Transgenic Plants—FDA Draft Federal Register Notice on Food Biotechnology', February 5, 1992, www.biointegrity.org

13 David Kessler, 'FDA Proposed Statement of Policy Clarifying the Regulation of Food Derived from Genetically Modified Plants—DECISION', March 20, 1992, www.biointegrity.org

14 James B. MacRae, Jr., Office of Management and Budget, to C. Boyden Gray, President Bush's White House counsel, 'FDA Food Biotechnology Policy', March 21, 1992, www.biointegrity.org

15 Eric Katz to John Gallivan, 'Food Biotechnology Policy Statement', March 27, 1992, www.biointegrity.org

16 Bill Lambrecht, *Dinner at the New Gene Café: How Genetic Engineering Is Changing What We Eat, How We Live, and the Global Politics of Food*, St. Martin's Press, New York, 2001, p. 322

17 'Speaker Hastert Calls for End of European Union's "Protectionist, Discriminatory Trade Policies"', US Newswire, March 26, 2003

18 James Maryanski to Dr. Bill Murray, Chairman of the Food Directorate, Canada, 'The safety assessment of foods and food ingredients developed through new biotechnology', October 23, 1991, www.biointegrity.org

19 Andrea Baillie, 'Suzuki Warns of Frankenstein Foods', *CP Wire*, October 18, 1999

20 'Expert Panel on the Future of Food Biotechnology', January, 2001, http://www.rsc.ca/foodbiotechnology/GMreportEN.pdf

21 Robert J. Scheuplein to the FDA Biotechnology Coordinator and others, 'Response to Calgene Amended Petition', October 27, 1993, www.biointegrity.org

22 Carl B. Johnson to Linda Kahl and others, 'Flavr Savr™ tomato; significance of pending DHEE question', December 7, 1993, www.biointegrity.org

23 Fred Hines to Linda Kahl, 'FLAVR SAVR Tomato' . . . 'Pathology Branch's Remarks to Calgene Inc.'s Response to FDA Letter of June 29, 1993', www.biointegrity.org

24 Arpad Pusztai, 'Genetically Modified Foods: Are They a Risk to Human/Animal Health?' June 2001

25 Murray Lumpkin to Bruce Burlington, 'The tomatoes that will eat Akron', December 17, 1992, www.biointegrity.org

26 Albert Sheldon to James Maryanski, Biotechnology Coordinator, 'Use of Kanamycin Resistance Markers in Tomatoes', March 30, 1993, www.biointegrity.org

27 Ricki Lewis, The Rise of Antibiotic-Resistant Infections FDA page', *FDA Consumer magazine*, September 1995, http://www.fda.gov/fdac/features/795_antibio.html

28 'Q&A: MRSA "superbugs"', *BBC News online*, December 13, 2002, http://news.bbc.co.uk/1/hi/health/2572841.stm

29 Code of Federal Regulations Title 21, Sec. 170.30(b)

30 Personal communication with James Turner

31 Michael Culbert, *Medical Armageddon*, C and C Communications, 1997

32 Marion Nestle, *Food Politics: How the Food Industry Influences Nutrition and Health*, University of California Press, Berkeley and Los Angeles, California, 2002

33 Craig Canine, 'Hear No Evil', *Eating Well*, July/August 1991

34 FDA Drug Review; Postapproval Risks 1976-85 United States General Accounting Office (GAO) GAO/PEMD-90-15, April 1990

35 Center for the Study of Drug Development, Boston, 1990

36 Michael Pollan, 'Playing God in the Garden', *The New York Times Sunday Magazine*, October 25, 1998

37 Debora MacKenzie, 'Unpalatable Truths', *New Scientist* Special Report: Living in a GM World, 1999

38 Laura Ticciati and Robin Ticciati, *Genetically Engineered Foods: Are They Safe? You Decide*. Keats Publishing, New Canaan, Connecticut, 1998, pp. 13-14

39 'US to keep a closer watch on genetically altered crops', *New York Times*, May 4, 2000

40 'US Rep. Kucinich terms new genetically engineered food regs "inadequate"', Kucinich press release, May 3, 2000

41 'Biofood rules mean few changes for companies', Reuters, May 3, 2000

42 Emily Gersema, 'FDA Opts against Further Biotech Review', Associated Press, June 17, 2003

43 Michael Grunwald, 'Monsanto hid decades of pollution in Alabama town', *Washington Post*, January 1, 2002, p. A01

44 Brian Tokar, 'Monsanto: A Checkered History', *The Ecologist*, Sep/Oct 1998, http://www.mindfully.org/Industry/Monsanto-Checkered-HistoryOct98.htm

45 'EPA Memo Says Monsanto Fabricated Dioxin Data', *The Milkweed*, February, 1990

46 Micah L. Sifry, 'Food Money', *The Nation* magazine, December 27, 1999, http://www.thirdworldtraveler.com/Corporations/FoodMoney.html

47 Robert Cohen testimony before FDA panel, December 2, 1999

48 Gwynne Dyer, 'Frankenstein Foods',*Globe and Mail*, February 20, 1999, http://www.netlink.de/gen/Zeitung/1999/990220d.htm

49 Paul Elias, 'Biotech lobbyists' clout grows in Washington', Associated Press, June 2, 2002

50 'Speaker Hastert Calls for End of European Union's "Protectionist, Discriminatory Trade Policies"', US Newswire, March 26, 2003
51 'Eye Opening Job Switches: GEMA Industry and US Gov.' The Edmonds Institute, The Third World Network and friends
52 Molly Ivans, 'White House '04 campaign 2004 early analysis', *Ft. Worth Star-Telegram*, February 9, 2001
53 Robert Cohen, 'Pelican Brief', www.notmilk.com
54 Bill Lambrecht, *Dinner at the New Gene Café*, pp. 197-198
55 Bill Lambrecht, *Dinner at the New Gene Café*, p. 139
56 Bill Lambrecht, 'Outgoing Secretary says Agency's Top Issue Is Genetically Modified Food', *St. Louis Post-Dispatch*, January 25, 2001
57 Hugh Warwick and Gundala Meziani, *Seeds of Doubt*, UK Soil Association, September 2002
58 'GE crops—increasingly isolated as awareness and rejection grow', Greenpeace International, briefing, March 2002
59 S. Branford, 'Sow resistant', *The Guardian*, April 17, 2002
60 Charles Benbrook, 'Premium Paid for Bt Corn Seed Improves Corporate Finances While Eroding Grower Profits', Benbrook Consulting Services, Sandpoint, Idaho, February 2002
61 UK Soil Association figure, consisting of $3–5 billion annually in extra farm subsidies, $2 billion in lost foreign trade and $1 billion cost of the StarLink accident
62 Bill Lambrecht, *Dinner at the New Gene Café*, p. 243
63 Sheldon Rampton and John Stauber, *Trust Us We're Experts*, Jeremy P. Tarcher/
Putnam, New York, 2001, p. 182
64 Adam Jasser, 'US says EU needs own Food and Drug Administration', Reuters, September 16, 1999

'Wisdom of the Rats'
1 Rick Weiss, 'Biotech Food Raises a Crop of Questions', *Washington Post*, August 15, 1999; p. A1

Chapter 6
1 Rajeev Syal, 'GM soya milk gives children herpes, senior surgeon tells the Government', *Sunday Telegraph*, August 1, 1999
2 Mark Townsend, 'Why soya is a hidden destroyer', *Daily Express*, March 12, 1999
3 Stephen R. Padgette and others, 'The Composition of Glyphosate-Tolerant Soybean Seeds Is Equivalent to That of Conventional Soybeans', *The Journal of Nutrition*, vol. 126, no. 4, April 1996 (The data was taken from the journal archives, as it had been omitted from the published study.)
4 Rick Weiss, 'Biotech Food Raises a Crop of Questions', *Washington Post*, August 15, 1999, p. A1
5 J. D. Nordlee and others, 'Identification of a brazil nut allergen in transgenic soybeans', *New England Journal of Medicine*, vol. 334, no. 11, 1996, p. 726

6   Louis J. Pribyl, 'Biotechnology Draft Document, 2/27/92', March 6, 1992, www.biointegrity.org

7   'Statement of Policy: Foods Derived from New Plant Varieties', Food and Drug Administration Docket No. 92N-0139

8   Personal communication with James Maryanski

9   GM-FREE Magazine, vol. 1, no. 3, August/September 1999

10 'Allergies to Transgenic Foods—Questions of Policy', Editorial New England Journal of Medicine, vol. 334, issue 11, 1996

11 SAP, 2000, no. 7, www.epa.gov/scipoly/sap/2000/february/foodal.pdf

12 Carl B. Johnson, Memo on the 'draft statement of policy 12/12/91', January 8, 1992

13 Arpad Pusztai, 'Genetically Modified Foods: Are They a Risk to Human/Animal Health?' June 2001, http://www.actionbioscience.org/biotech/pusztai.html

14 Personal conversation with Arpad Pusztai

15 J. M. Wal, 'Strategies for Assessment and Identification of Allergenicity in (Novel) Foods', International Dairy Journal, vol. 8, 1998, p. 422

16 UK Royal Society, 'Genetically modified plants for food use and human health—an update', February 2002, p. 8, www.royalsoc.ac.uk

17 Bill Freese, 'A Critique of the EPA's Decision to Re-Register Bt Crops and an Examination of the Potential Allergenicity of Bt Proteins', Submission by Friends of the Earth to the EPA, December 9, 2001

18 'Life-Threatening Food?' CBS News, May 17, 2001, http://www.cbsnews.com/now/story/0,1597,291992-412,00.shtml

19 Jonathan Bernstein, and others, 'Clinical and laboratory investigation of allergy to genetically modified foods', Environmental Health Prospectus, vol. 111, 2003, pp. 1114-1121

20 Marc Kaufman, 'Biotech Corn Is Test Case for Industry Engineered Food's Future Hinges on Allergy Study', Washington Post, March 19, 2001

21 Alan Rulis, Center for Food Safety and Applied Nutrition, to Sally Van Wert, AgrEvo USA Company, May 29, 1998, www.cfsan.fda.gov/~acrobat2/bnfL041.pdf

22 William Ryberg, 'Growers of biotech corn say they weren't warned: StarLink tags appear to indicate it's suitable for human food products', Des Moines Register, October 25, 2000.

23 Bill Freese, 'The StarLink Affair, Submission by Friends of the Earth to the FIFRA Scientific Advisory Panel considering Assessment of Additional Scientific Information Concerning StarLink Corn', July 17-19, 2001

24 Marc Kaufman, 'Genetically Engineered Corn Cleared in 17 Food Reactions', Washington Post, June 14, 2001

25 'Assessment of Additional Scientific Information Concerning StarLink Corn', FIFRA Scientific Advisory Panel Report No. 2001-09, July 2001

26 'Biotech Firm Executive Says Genetically Engineered Corn Is Here to Stay', Knight Ridder/Tribune, March 20, 2001

27 Richard B. Raybourne, 'Development and Use of a Method for Detection of IgE Antibodies to Cry9C', FDA, June 13, 2001

28 'Mammalian Toxicity Assessment Guidelines for Protein Plant Pesticides', FIFRA Scientific Advisory Panel Report No. 2000-03B, September 28, 2000
29 FIFRA Scientific Advisory Panel (SAP), Open Meeting, July 17, 2001
30 Masaharu Kawata, 'Dr. M. Kawata: Questions CDC and EPA testing of Starlink and allergies', *Renu Namjoshi*, June 28, 2001
31 Marc Kaufman, 'EPA Rejects Biotech Corn as Human Food; Federal Tests Do Not Eliminate Possibility That It Could Cause Allergic Reactions, Agency Told', *Washington Post*, July 28, 2001, p. A02
32 Marc Kaufmann 'Biotech Corn Is Test Case For Industry', *Washington Post*, March 19, 2001
33 Michael Hansen, 'Bt Crops: Inadequate Testing', Lecture delivered at Universidad Autonoma, Chapingo, Mexico, August 2, 2002
34 I. Bernstein and others, 'Immune responses in farm workers after exposure to Bacillus thuringiensis pesticides', *Environmental Health Perspectives*, vol. 107, 1999, pp. 575-582
35 Michael Hansen testimony: Possible Human Health Hazards of Genetically Engineered Bt Crops Presented to the EPA Science Advisory Panel, Arlington, VA, October 20, 2000
36 R. Vázquez, and others, *'Bacillus thuringiensis* Cry1Ac protoxin is a potent systemic and mucosal adjuvant', *Scandinavian Journal of Immunology*, vol. 49, 1999, 578-584
37 Joe Cummins, *'Bacillus thuringiensis* (Bt) toxin causes Allergy', March 5, 2000, www.purefood.org/ge/btcomments.cfm
38 Joel E. Ream for Monsanto, 'Assessment of the *In vitro* Digestive Fate of *Bacillus thuringiensis* subsp. kurstaki HD-1 Protein', March 23, 1994, Unpublished study submitted to the EPA, EPA MRID No. 434392-01
39 Bill Freese, 'A Critique of the EPA's Decision to Re-Register Bt Crops and an Examination of the Potential Allergenicity of Bt Proteins', adapted from Friends of the Earth submission to the EPA, Dec. 9, 2001, www.foe.org/safefood/comments.pdf
40 'Evaluation of Allergenicity of Genetically Modified Foods', Report of a Joint FAO/WHO Expert Consultation on Allergenicity of Foods Derived from Biotechnology, January 22-25, 2001
41 H. Noteborn, 'Assessment of the Stability to Digestion and Bioavailability of the LYS Mutant Cry9C Protein from *Bacillus thuringiensis* serovar tolworthi', unpublished study submitted to the EPA by AgrEvo, EPA MRID No. 447343-05, 1998
42 Noteborn and others, 'Safety assessment of the *Bacillus thuringiensis* insecticidal crystal protein CRYIA(b) expressed in transgenic tomatoes', in Engel, et al. (eds.), *American Chemical Society Symposium* Series 605, Washington, D.C., pp. 134-47, 1995
43 N. H. Fares and A. K El-Sayed, 'Fine structural changes in the ileum of mice fed on endotoxin-treated potatoes and transgenic potatoes', *Natural Toxins*, vol. 6, 1998, pp. 219-233
44 Gendel, 'The use of amino acid sequence alignments to assess potential allergenicity of proteins used in genetically modified foods', *Advances in Food and Nutrition Research*, vol. 42, 1998, pp. 45-62

45 G. A. Kleter and A. A. C. M. Peijnenburg, 'Screening of transgenic proteins expressed in transgenic food crops for the presence of short amino acid sequences indentical to potential, IgE-binding linear epitopes of allergens', *BMC Structural Biology*, vol. 2, 2002, p. 8-19

'Missing Chickens'
1 'GM safety tests "flawed"', *BBC News*, April 27, 2002
  http://news.bbc.co.uk/hi/ english/sci/tech/newsid_1954000/1954408.stm

Chapter 7
1 Personal communication with Jane Akre
2 Robert Cohen, *Milk, the Deadly Poison*, Argus Publishing, Englewood Cliffs, New Jersey, 1998
3 'Can two reporters take on Murdoch and win?' *The Independent* (London), September 14, 1999
4 Samuel Epstein and Pete Hardin, 'Confidential Monsanto Research Files Dispute Many bGH Safety Claims', *The Milkweed*, January 1990
5 'Milk, rBGH, and Cancer', *Rachel's Environment and Health Weekly*, no. 593, April 9, 1998
6 BGH Bulletin, Target Television Enterprises Inc., http://www.foxbghsuit.com/
7 Personal communication with Steve Wilson
8 'Growth Hormones Would Endanger Milk', Op-ed article, *Los Angeles Times*, July 27, 1989
9 Sheldon Rampton and John Stauber, *Trust Us We're Experts*, Jeremy P. Tarcher/Putnam, New York, 2001, p. 164
10 Samuel Epstein, 'FDA Is Ignoring Dangers of Bovine Growth Hormone', Letter to the editor, *Austin American Statesman*, June 2, 1990
11 'US Newspapers Present Biased View of Biotech', News Release, April 29, 2002, http://www.organicconsumers.org/corp/usnewsbias043002.cfm
12 Mae-Wan Ho and Jonathan Matthews, 'Suppressing Dissent in Science with GM Foods', http://www.mercola.com/2001/mar/14/gm_foods.htm
13 Personal communication with Bill Lashmett
14 'Who's Afraid of Monsanto? Britain's Best-Loved Newsagents Bend to History of Intimidation', press release, *Ecologist*, October 26, 1998
15 'Recycled Ecology', *SchNews*, Issue 185, October 2, 1998, http://www.schnews.org.uk/archive/news185.htm
16 Paul Brown, 'Printers pulp Monsanto edition of Ecologist', *The Guardian*, September 29, 1998
17 'The Monsanto Files: Can We Survive Genetic Engineering?' *Ecologist*, vol. 28 no. 5, September/October 1998
18 Center for Ethics and Toxics, www.cetos.org
19 Charles Benbrook, 'Evidence of the Magnitude and Consequences of the Roundup Ready Soybean Yield Drag from University-Based Varietal Trials in 1998', Ag BioTech InfoNet Technical Paper Number 1, July 13, 1999, http://www.biotech-info.net/ RR_yield_drag_98.pdf

20 Marc Lappé and Britt Bailey, *Against the Grain: Biotechnology and the Corporate Takeover of Your Food*, Common Courage Press, Monroe, Maine, 1998, p. 76

21 Bill Lambrecht, *Dinner at the New Gene Café: How Genetic Engineering Is Changing What We Eat, How We Live, and the Global Politics of Food*, St. Martin's Press, New York, 2001, p. 77

22 Becky Goldburg, 'Preliminary Research Results Presented during the Monarch Research Symposium', Environmental Defense Fund, New York, November 2, 1999, http://www.biotech-info.net/symposium_report.html

23 Carol Kaesuk Yoon, 'No Consensus on the Effects of Engineering on Corn Crops', *New York Times*, November 4, 1999

24 Sheldon Rampton and John Stauber, *Trust Us We're Experts*, Jeremy P. Tarcher/
Putnam, New York, 2001, pp. 186-187

25 Carol Kaesuk Yoon, 'What's Next for Biotech Crops? Questions', *New York Times*, December 19, 2000

26 Bill Lambrecht, *Dinner at the New Gene Café*, p. 267

27 Michael Pollan, 'The Great Yellow Hype', *New York Times*, March 4, 2001, section 6, p. 15

28 'GE rice is fool's gold', Greenpeace, http://archive.greenpeace.org/~geneng/highlights/food/goldenrice.htm

29 Greenpeace demands false biotech advertising be removed from TV, Letter, February 9, 2001

30 Opinion piece about Golden Rice by Benedikt Haerlin, http://archive.greenpeace.org/~geneng/highlights/food/benny.htm

31 'Grains of Delusion', Jointly published by BIOTHAI (Thailand), CEDAC (Cambodia), DRCSC (India), GRAIN, MASIPAG (Philippines), PAN-Indonesia and UBINIG (Bangladesh), February 2001, www.grain.org/publications/delusion-en.cfm

32 Vitamin Angel Alliance, http://www.vitaminangelalliance.com/vitamina.html

33 Kristi Coale, 'Mutant food', *Salon*, January 12, 2000, http://archive.salon.com/news/feature/2000/01/12/food/

34 Robert S. Greenberger, 'Motley Group Pushes for FDA Labels On Biofoods', *The Wall Street Journal*, August 18, 1999

35 Rick Weiss, 'Next Food Fight Brewing Is over Listing Genes on Labels', *Washington Post*, August 15, 1999, p. A17

36 Personal communication with Steven M. Druker

37 Steven M. Druker, http://www.bio-integrity.org/report-on-lawsuit.htm

38 James Walsh, 'Brave New Farm', *Time*, January, 1999

39 Personal communication with Laura Ticciati

40 http://www.voteyeson27.com

41 William K. Jaeger, 'Economic Issues and Oregon Ballot Measure 27: labelling of Genetically Modified Foods', Oregon State University Extension Service, October 2002

42 Bill Lambrecht, *Dinner at the New Gene Café*, p. 251

43 Steven Druker, letter to Governor John A. Kitzhaber, M.D., October 10, 2002, http://ngin.tripod.com/121002b.htm

44 Pauletter Pyle, Andy Anderson, and Terry Witt, Letter on behalf of Oregonians for Food and Shelter, September 12, 2002
45 Michelle Cole, 'Bill would restrict food-label regulation', *Oregonian*, April 12, 2003
46 John Vidal, 'Biotech food giant wields power in Washington', *The Guardian*, February 18, 1999
47 All quotes by Ignacio Chapela and those attributed to Fernando Ortiz Monasterio are based on personal commuincations with Ignacio Chapela, unless otherwise specified.
48 Paul Brown, *The Guardian*, April 19, 2002 http://www.guardian.co.uk/ gmde-bate/Story/0,2763,686955,00.html
49 'Seeds of Trouble', *BBC*, January 7 and 14, 2003
50 Mark Henderson, 'Attack on safety of GM crops was unfounded', *London Times*, April 5, 2002
51 Paul Elias, 'Corn Study Spurs Debate over Corporate Meddling in Academia', Associated Press, April 18, 2002
52 C. K. Yoon, 'Journal Raises Doubts on Biotech Study', *New York Times*, April 5, 2002.
53 George Monbiot, 'The Fake Persuaders: Corporations are Inventing People to Rubbish Their Opponents on the Internet', *The Guardian*, May 14, 2002, www.monbiot.com
54 George Monbiot, 'The covert biotech war', *The Guardian*, November 19, 2002, www.monbiot.com
55 Paul Brown, 'Mexico's vital gene reservoir polluted by modified maize', *The Guardian*, April 19, 2002
56 Charles Clover, '"Worst ever" GM Crop Invasion', *Daily Telegraph*, UK, April 19, 2002
57 Ronnie Cummins, 'Exposing Biotech's Big Lies', *BioDemocracy News,* no. 39, May 2002, www.organicconsumers.org

'Changed Mice'
1 Hinze Hogendoorn, http://www.talk2000.nl/mice/talk-Extended.htm
2 Steve Sprinkel, 'When the Corn Hits the Fan', *Acres*, U.S.A., September 18, 1999
3 Gundula Meziani and Hugh Warwick, *'Seeds of Doubt'*, Soil Association (UK), September 17, 2002

Chapter 8
1 'A different kind of school lunch', *Pure Facts*, Feingold® Association of the United States, October 2002, http://www.feingold.org/newsletter1.html
2 Barbara Reed Stitt, *Food and behaviour, A Natural Connection*, Natural Press, Manitowoc, Wisconsin, 1997
3 Personal communication with Greg Britthauer
4 Personal communication with Barbara Reed Stitt
5 Personal communication with sister Luigi Frigo

6 Bill Scanlon, 'Food has big effects on health', *Rocky Mountain News*, December 13, 2002
7 Tara Womersley, 'Food Colourings Cause 1 in 4 Temper Tantrums, Claims British Study', *The Scotsman*, October 25, 2002
8 Ronnie Cummins, 'Frankencorn Fight: Cautionary Tales', *BioDemocracy News*, no. 37, May 2002, www.organicconsumers.org
9 Genetic Food Alert (UK), Press Release, September 2002, www.connectotel.com/gmfood/gf150902.txt
10 Ken Roseboro, 'Manufacturers face GMO challenges with minor ingredients', *Non-GMO Source*, September 2002
11 Genetically Engineered Organisms, Public Issues Project, http://www.geo-pie.cornell.edu//gmo.html
12 John Robbins, *The Food Revolution: How Your Diet Can Help Save Your Life and Our World*, Conari Press, Berkeley, California, 2001, p. 346

'A Strange Deal'
1 'EU's Nielson blasts US 'lies' in GM food row', Reuters, January 20, 2003, http://ngin.tripod.com/200103d.htm

Chapter 9
1 Bill Lambrecht, *Dinner at the New Gene Café: How Genetic Engineering Is Changing What We Eat, How We Live, and the Global Politics of Food*, St. Martin's Press, New York, 2001, p. 293
2 'Myths About World Hunger: Myth 1: Not Enough Food to Go Around', Stop Hunger Now, http://www.stophungernow.org/hunger_facts.html
3 'Agriculture: Towards 2015/30', Food and Agriculture Organization of the United Nations, July 24, 2000
4 Robert Shapiro, 'The Welcome Tension of Technology: The Need for Dialogue about Agricultural Biotechnology', Center for the Study of American Business, CEO Series 37, February 2000
5 John Robbins, *The Food Revolution: How Your Diet Can Help Save Your Life and Our World*, Conari Press, Berkeley, California 2001, p. 341
6 Dorothy Mclaughlin, 'Fooling with Nature, Silent Spring Revisited', *PBS*, http://www.pbs.org/wgbh/pages/frontline/shows/nature/disrupt/sspring.html
7 Dominic Rushe. 'Major UK Grocery Chain Bans Aspartame over Brain Tumour Concerns', *The Sunday Times* (London), October 25, 1999, http://www.sunday-times.co.uk
8 Carol Baxter, 'Reading, 'Riting and rBGH', Healthwell, http://www.healthwell.com/delicious-online/D_Backs/Sep_97/gl.cfm?path=hw

Epilogue
1 Michael Meacher, 'Are GM crops safe? Who can say? Not Blair', The *Independent*, June 22, 2003
2 'The Case for A GM-Free Sustainable World', Independent Science Panel, May 10, 2003, www.indsp.org

Index

About the Author

Jeffrey M. Smith has been active in the field of genetically modified (GM) foods for nearly a decade. He worked for non-profit and political groups on the issue and in 1998, ran for US Congress to raise public awareness of the health and environmental impacts. Smith proposed legislation to remove GM food from school meals, to protect children, who are most at risk from the potential health effects, and legislation to help protect farmers from cross-pollination by GM crops. Later, he was vice president of marketing for a GMO detection laboratory.

Smith has lectured widely, spoken at conferences, and has been quoted in articles around the world. Prior to working in this field, he was a writer, educator, and public speaker for non-profit groups, advancing the causes of health, environment, and personal development. This book, researched and written after he left the industry, combines Smith's passion for these causes with his extensive knowledge of the risks and cover-ups behind genetically modified foods.

Smith is the founder and director of the Institute for Responsible Technology. He has a master's degree in business administration and lives with his wife in Iowa.

Author's Note

I do not wish to disparage the many hardworking, competent people at the FDA and other agencies of the US government. As demonstrated by the FDA scientists whose recommendations for mandatory safety testing were ignored, many conscientious government employees are forced to work under policies that are not based on sound science and not in the best interest of the public.

How to order more copies of this book

You can buy Green Books at bookshops and green retailers. If you wish to buy more copies of this book, you can also order direct from us using the form below.

Name _____

Address _____

_____ Postcode _____ Phone _____

	Quantity	Amount £
Copies of *Seeds of Deception* @ £9.95		
6 copies or more @ £6.50 each		
Subtotal:	_____	_____
UK: Post & Packing on all orders		£1.50
Overseas: add £2.00 for 1st book, £1 for each additional book		_____
Order Total:		£ _____

Payment information:

☐ By cheque: make payable to Green Books, and send with this form to the above address.

☐ By credit card: fill out the information below.

Payment by credit card:

Credit Card: Visa ☐ MasterCard ☐ Switch ☐ No. _____

Expiry Date _____ Start Date _____ Issue no. (Switch only) ___ Signature _____

Name as it appears on the card _____

Send this form to the address below. If you have any questions, please contact us.

Green Books,
Foxhole, Dartington,
Totnes, Devon TQ9 6EB
Tel: 01803 863260
Fax: 01803 863843
email: sales@greenbooks.co.uk